Common Core Standards

Algebra II Practice Tests

Common Core State Standards ® Introduction & Curriculum

The Common Core State Standards provide a consistent, clear curriculum about what students are expected to learn, so teachers and parents know what they need to do to help. Various states and school districts offer tests that measure student proficiency. These practice tests are based on the Common Core Standards curriculum at http://www.corestandards.org/

Contents

Algebra 2-Questions: Section 1

1. What is the complete solution to the equation $|4 - 8x| = 12$?

a. $x = 1; x = -2$

b. $x = -1; x = 2$

c. $x = 1; x = 2$

d. $x = -1; x = -2$

2. What are the possible values of x in $|15 - 3x| = 6$?

a. $7 > x > -3$

b. $-7 < x < 3$

c. $x = 3$ or $x = 7$

d. $x = -3$ or $x = -7$

3. Twenty percent of the people at the concert bought their tickets during the pre-sale. If 1500 bought their tickets during the regular sale of tickets, how many attended the concert?

a. 1875

b. 1500

c. 1125

d. 1900

4. What is the solution to the system of equations shown below?

$$\begin{cases} 2x + 3y - z = -3 \\ x + 2y = 0 \\ x - 2y + z = -2 \end{cases}$$

a. No solutions

b. Infinitely many solutions

c. (-2,1,2)

d. $(\frac{10}{7}, \frac{5}{7}, -2)$

5. A store manager bought 15 packages of socks. Some packages contain 8 pairs of socks and the rest contain 10 pairs of socks. There were 132 pairs in all. How many packages of 10 pairs of socks did the manager buy?

a. 6

b. 7

c. 9

d. 10

6. What system of inequalities *best* represents the graph shown below?

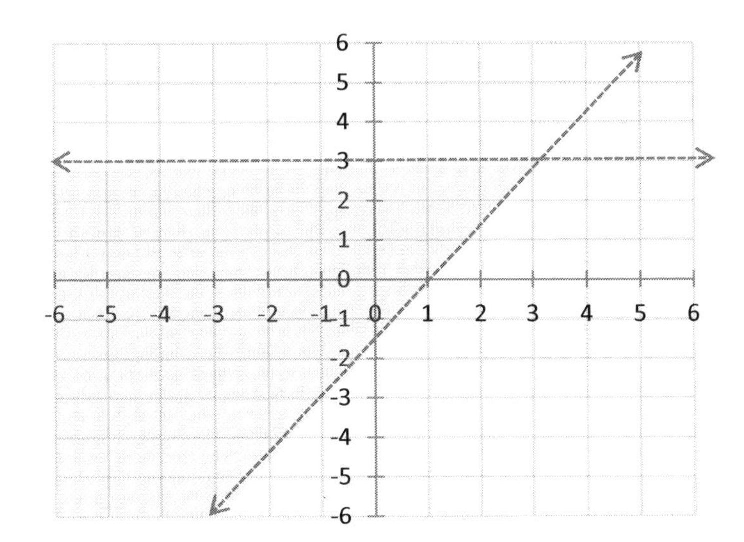

a. $y > 3$ and $y < x - 1$

b. $y > 3$ and $y > x - 1$

c. $y < 3$ and $y < x - 1$

d. $y < 3$ and $y > x - 1$

7. Which point lies in the solution set for the system $\begin{cases} 3y + 2x > -6 \\ -3y + x \geq -6 \end{cases}$?

a. (0,-4)

b. (1,2)

c. (-4,2)

d. (1,-3)

8. Which system of linear inequalities is represented by this graph?

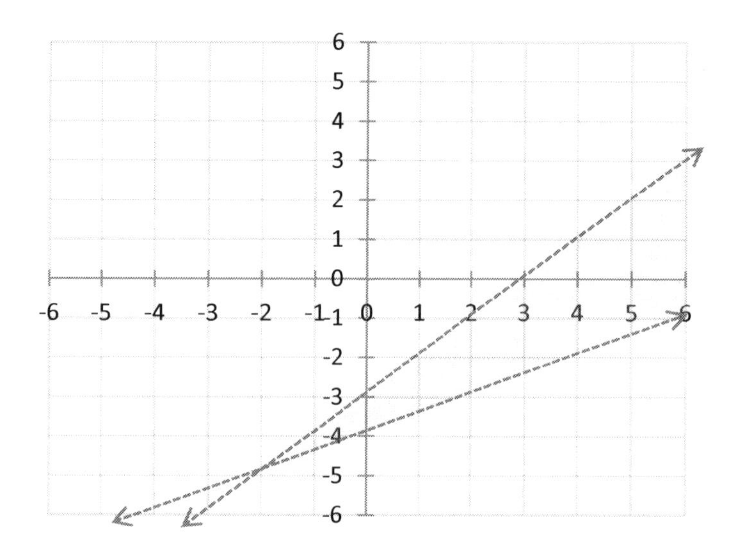

a. $\begin{cases} y < \frac{1}{3}x + 4 \\ y < x + 3 \end{cases}$

b. $\begin{cases} y > \frac{1}{3}x - 4 \\ y < x - 3 \end{cases}$

c. $\begin{cases} y < \frac{1}{3}x - 4 \\ y > x - 3 \end{cases}$

d. $\begin{cases} y < \frac{1}{3}x + 4 \\ y > x + 3 \end{cases}$

9. Which polynomial represents $(2x^2 - 2x - 2)(x + 4)$?

a. $2x^3 - 6x^2 - 10x - 8$

b. $2x^3 - 6x^2 - 10x + 8$

c. $2x^3 + 6x^2 - 10x - 8$

d. $2x^3 - 6x^2 + 10x - 8$

10. $x-4\overline{)3x^3-2x^2+5x+2}$

a. $3x^2+10x+45+\frac{182}{x-4}$

b. $3x^2-10x+45-\frac{182}{x-4}$

c. $3x^2+10x-45+\frac{182}{x-4}$

d. $3x^2-10x-45-\frac{182}{x-4}$

11. $(-3x^2+5x+2)-3(2x^2-2x+1)=$

a. $3x^2-1$

b. $-9x^2-1$

c. $3x^2+11x-1$

d. $-9x^2+11x-1$

12. Which expression is equivalent to $(3y^2-1)(3y+1)$?

a. $9y^2-1$

b. $9y^3-1$

c. $9y^3+3y^2-3y-1$

d. $9y^3+3y^2+3y-1$

13. What is the volume of the figure below?

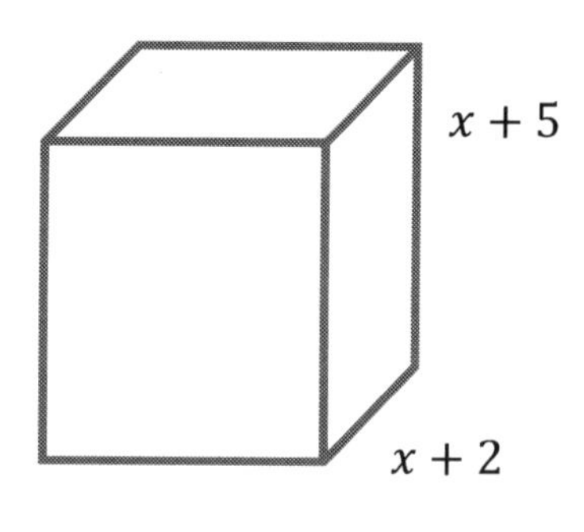

a. $x^3 + 11x^2 + 31x + 30$
b. $x^3 + 10x^2 + 31x + 30$
c. $x^3 + 11x^2 + 21x + 30$
d. $x^3 + 10x^2 + 21x + 30$

14. $18a^3 - 3a^2c + c^3 =$

a. $(3a + c)(6a^2 - 3ac + c^2)$
b. $(3a - c)(6a^2 - 3ac + c^2)$
c. $(3a - c)(3a^2 - 3ac + c^2)$
d. $(3a + c)(3a + c)(3a + c)$

15. The total area of a rectangle is $25x^4 - 16y^2$. Which factors could represent the length times width?

a. $(5x^2 - 4y)(5x^2 + 4y)$
b. $(5x^2 - 4y)(5x^2 - 4y)$
c. $(5x + 4y)(5x + 4y)$
d. $(5x + 4y)(5x - 4y)$

16. Which product of factors is equivalent to $(x-1)^2 + y^2$?

a. $(x-1+y)^2$
b. $(x+1-y)^2$
c. $(x-1+y)(x-1+y)$
d. $(x+1+y)(x+1-y)$

17. $\frac{x-4}{x-3} + \frac{8}{x^2-x-6} =$

a. $\frac{x^2+2x+6}{x^2-x-6}$
b. $\frac{2x-6}{x^2-x-6}$
c. $\frac{x^2-2x-6}{x^2-x-6}$
d. $\frac{x^2-2x}{x^2-x-6}$

18. Which is a simplified form of $\frac{4ab^2c^{-1}}{(a^{-2}b^3c)^3}$?

a. $\frac{4}{b^7c}$
b. $\frac{4a^7}{b^7c^2}$
c. $\frac{4}{a^7b^7c^2}$
d. $\frac{4ac}{b^7}$

19. $\frac{x^2+5x}{x^2+2x} \cdot \frac{x^2-4}{x^2-7x+10} =$

a. $\frac{x+5}{x-5}$

b. x

c. $x + 5$

d. 5

20. If $i = \sqrt{-1}$, which point shows the location of $-2 + i$ on the plane?

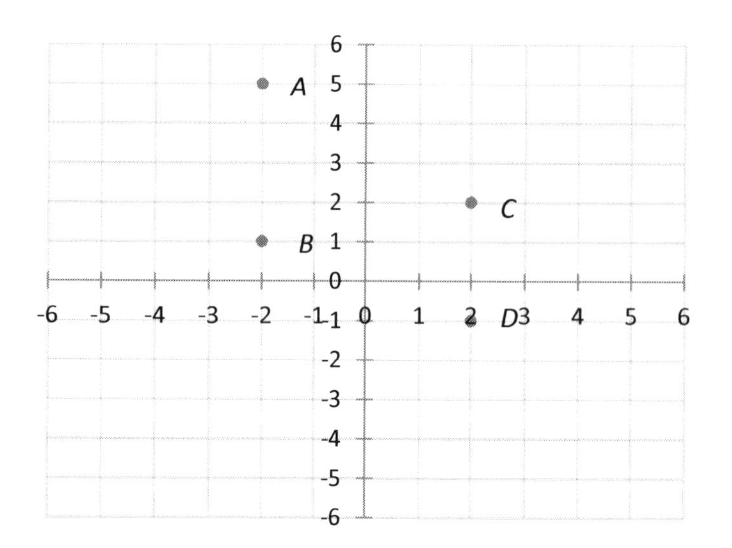

a. Point A

b. Point B

c. Point C

d. Point D

21. If $i = \sqrt{-1}$, what is the value of i^5?

a. i

b. $-i$

c. 1

d. -1

22. Which of the following complex numbers is represented by the point on the graph below?

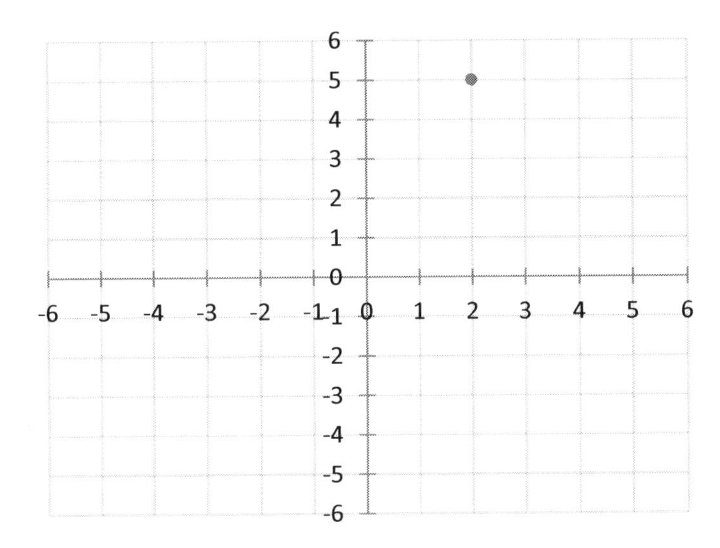

a. $5 + 2i$
b. $5 - 2i$
c. $2 + 5i$
d. $2 - 5i$

23. If $i = \sqrt{-1}$, then $3i(5i) =$

a. 15

b. -15

c.30

d. -30

24. What is an equivalent form of $\frac{2}{5+i}$?

a.$\frac{10-i}{13}$

b. $\frac{26-i}{13}$

c. $\frac{13-i}{13}$

d. $\frac{5-i}{13}$

25. What is the product of the complex numbers $(4 + i)$ and $(4 - i)$?

a. $16 - i$

b. $8 - 4i$

c. 17

d. 15

26. What are to the solutions to the equation $2x^2 - x + 3 = 0$?

a. $x = \frac{1}{4}; x = 1$

b. $x = \frac{1}{4} + i; x = \frac{1}{4} - i$

c. $x = \frac{1}{4}; x = \frac{23}{4}$

d. $x = \frac{1}{4} + \frac{\sqrt{23}}{4}i; x = \frac{1}{4} - \frac{\sqrt{23}}{4}i$

27. Mark is solving the equation $x^2 - 10x = 12$ by completing the square. What number should be added to both sides of the equation to complete the square?

a. 5

b. 10

c. 15

d. 25

28. What are the solutions to the equation $2 + \frac{4}{x^2} = -\frac{1}{x}$?

a. $x = \frac{1}{4} + \frac{\sqrt{31}}{4}; x = \frac{1}{4} - \frac{\sqrt{31}}{4}$

b. $x = -\frac{1}{4} + \frac{\sqrt{31}}{4}i; x = -\frac{1}{4} - \frac{\sqrt{31}}{4}i$

c. $x = -\frac{1}{4}; x = \frac{31}{4}$

d. $x = \frac{1}{4} + \frac{\sqrt{31}}{4}i; x = \frac{1}{4} - \frac{\sqrt{31}}{4}i$

29. Which of the following sentences is true about the graphs of $y = 2(x - 4)^2 + 3$ and $y = 2(x + 4)^2 + 3$?

a. One graph has a vertex that is a maximum, while the other graph has a vertex that is a minimum.
b. Their vertices are maximums.
c. The graphs have different shapes with different vertices.
d. The graphs have the same shape with different vertices.

30. Which of the following *most* accurately describes the translation of the graph $y = (x + 2)^2 - 1$ to the graph of $y = (x - 2)^2 + 3$?

a. up 4 and 4 to the right
b. down 3 and 2 to the left
c. up 2 and 2 to the right
d. down 1 and 4 to the left

31. What are the x-intercepts of the graph of $y = 6x^2 + 4x - 2$?

a. $\frac{1}{2}$ and $-\frac{2}{3}$

b. $-\frac{1}{2}$ and $\frac{2}{3}$

c. -1 and $\frac{1}{3}$

d. 1 and $-\frac{1}{3}$

32. Which is the graph of $y = (x - 2)^2 + 4$?

a.

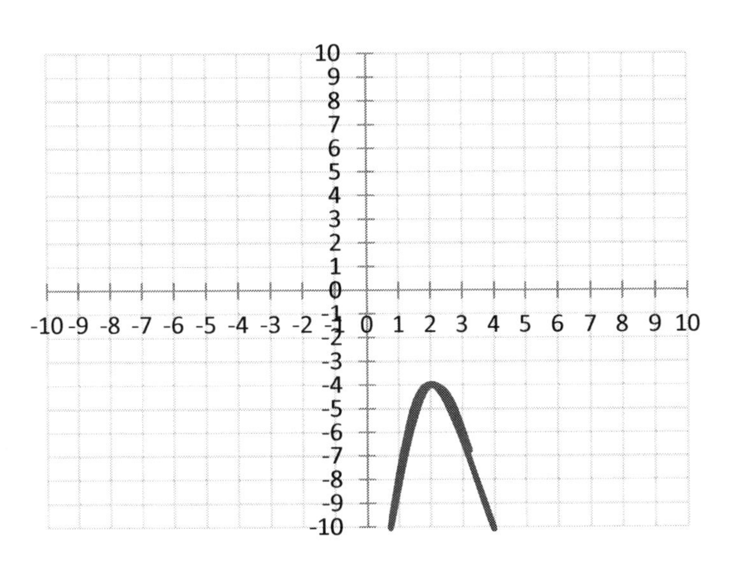

b.

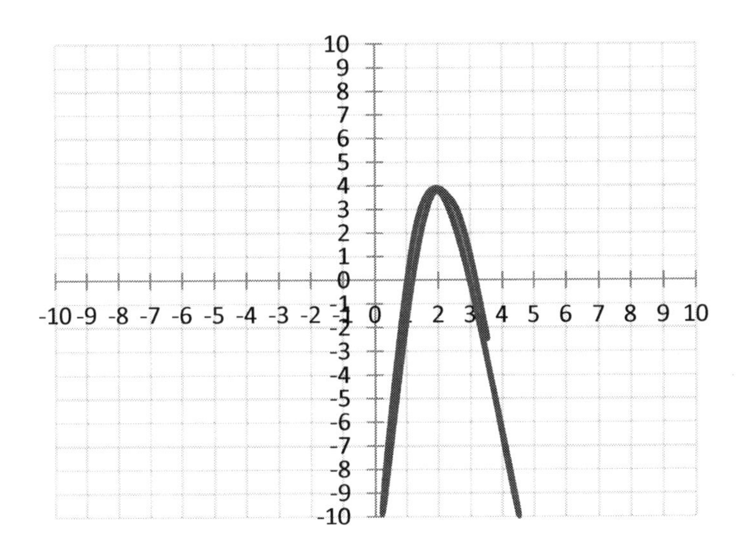

c.

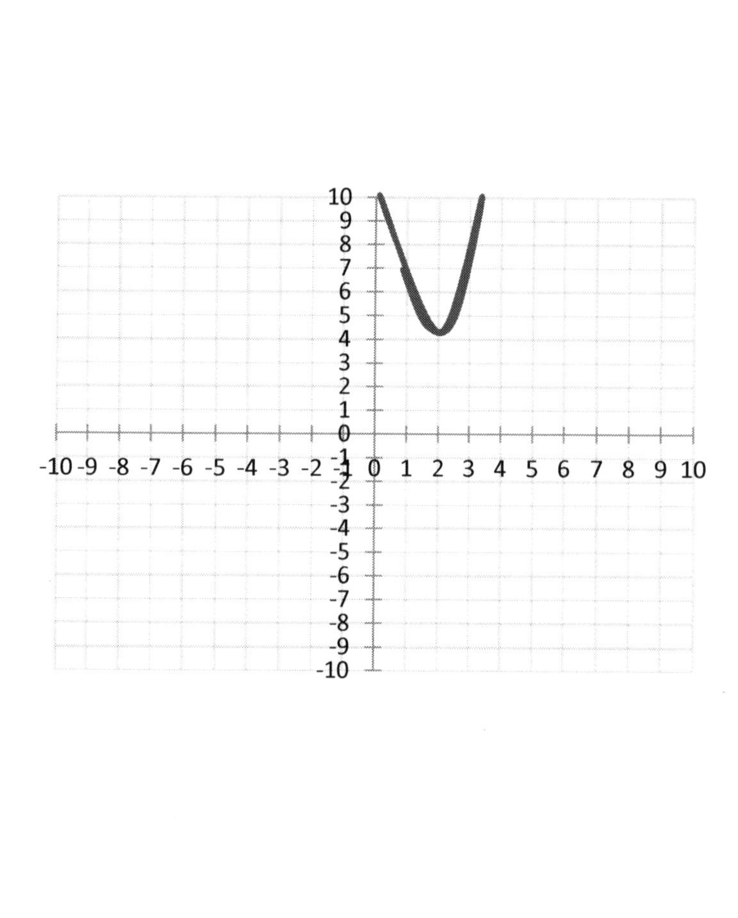

10
9
8
7
6
5
4
3
2
1
0
-1
-2
-3
-4
-5
-6
-7
-8
-9
-10
-10 -9 -8 -7 -6 -5 -4 -3 -2 -1 0 1 2 3 4 5 6 7 8 9 10

d.

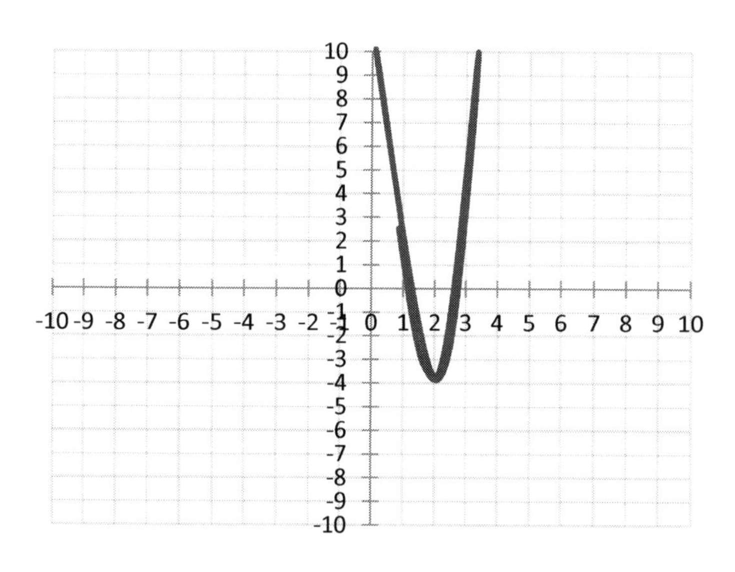

10
9
8
7
6
5
4
3
2
1
0
-1
-2
-3
-4
-5
-6
-7
-8
-9
-10
-10 -9 -8 -7 -6 -5 -4 -3 -2 -1 0 1 2 3 4 5 6 7 8 9 10

33. Which ordered pair is the vertex of $f(x) = x^2 - 10x + 25$?

a. $(1,10)$

b. $(15,4)$

c. $(2,3)$

d. $(5,0)$

34. Which of the following equations best represents a hyperbola?

a. $4px = y^2$

b. $\frac{(x-h)^2}{a^2} - \frac{(y-k)^2}{b^2} = 1$

c. $x^2 + y^2 = r^2$

d. $\frac{4}{3}\pi r^3$

35. The graph of $\left(\frac{x}{3}\right)^2 - \left(\frac{y}{2}\right)^2 = 1$ is a hyperbola. Which set of equations of represents the asymptotes of the hyperbola's graph?

a. $y = \frac{1}{2}x; y = -\frac{1}{2}x$

b. $y = \frac{1}{3}x; y = -\frac{1}{3}x$

c. $y = \frac{3}{2}x; y = -\frac{3}{2}x$

d. $y = \frac{2}{3}x; y = -\frac{2}{3}x$

36. $2x^2 - 7y^2 - 12x - 14y - 4 = 0$

What is the standard form of the equation of the conic given above?

a. $\frac{(x-3)^2}{7} - \frac{(y+1)^2}{2} = 1$

b. $\frac{(y-4)^2}{7} - \frac{(x+3)^2}{2} = 1$

c. $\frac{(x+7)^2}{3} - \frac{(y-2)^2}{6} = 1$

d. $\frac{(y+6)^2}{5} - \frac{(x-4)^2}{8} = 1$

37. What is the solution to the equation $3^x = 11$?

a. $x = \log_{10} 3 + \log_{10} 11$

b. $x = 8$

c. $x = \log_{10} 8$

d. $x = \frac{\log_{10} 11}{\log_{10} 3}$

$\log_3 11 = X$

38. If $\log_{10} x = -3$, what is the value of x?

a. $\frac{1}{1000}$

b. $\frac{1}{10}$

c. 10

d. $\sqrt{\frac{1}{100}}$

$10^{-3} = X$

$10 \times 10 \times 10 \cdot \frac{1}{100}$

39. Which equation is equivalent to $\log_3 \frac{1}{6} = x$?

a. $3^{\frac{1}{6}} = x$

b. $\frac{1^3}{6} = x^3$

c. $3^x = \frac{1}{6}$

d. $\left(\frac{1}{6}\right)^3 = x$

40. Which is the first *incorrect* step in simplifying $\log_5 \frac{5}{125}$?

Step 1: $\log_5 \frac{5}{125} = \log_5 5 - \log_5 125$

Step 2: $= 1 - 3$

Step 3: $= -2$

a. Step 1
b. Step 2
c. Step 3
d. Each step is correct.

41. Taylor, Jordan, Zac, and Avery each worked the same math problem at the chalkboard. Each student's work is shown below. Their teacher said that while two of them had the correct answer, only one of them had arrived at the correct conclusion using correct steps.

Taylor's Work

$x^4x^{-6} = \frac{x^4}{x^6}$

$= x^2; x \neq 0$

Zac's Work

$x^4x^{-6} = \frac{x^4}{x^{-6}}$

$= x^{10}; x \neq 0$

Jordan's Work

$x^4x^{-6} = \frac{x^4}{x^6}$

$= \frac{1}{x^2}; x \neq 0$

Avery's Work

$x^4x^{-6} = \frac{x^4}{x^{-6}}$

$= x^{-2}; x \neq 0$

a. Taylor

b. Jordan

c. Zac

d. Avery

42. A student showed the following steps in her solution of the equation below, but her answer was not correct.

$$\log_3(x+7) + \log_3 2 = \log_3 20$$

Step 1: $\mathbf{\log_3(x+7)(2) = 20}$

Step 2: $\mathbf{\log_3(2x+14) = 20}$

Step 3: $\mathbf{2x = 34}$

Step 4: $\mathbf{x = 17}$

In which step did she make her first error?

a. Step 1

b. Step 2

c. Step 3

d. Step 4

43. A certain car model loses its value over time according to the equation $y = A\left(\frac{1}{4}\right)^{\frac{t}{200}}$, where A= the price the car was initially and t =time in days. If 20,000 was the initial price, how much will the price of the car depreciate in 400 days?

a. 625

b. 1250

c. 2500

d. 5000

44. Mold spores in a contained unit are growing exponentially with time, as shown in the table below.

Mold Spores Growth

Day	Mold
0	50
1	100
2	200

Which of the following equations expressing the number of mold spores, y, present at any time, t?

a. $y = (50) \cdot (2)^t$
b. $y = t^2$
c. $y = 2^t + 50$
d. $y = (100) \cdot (2)^t$

45. If the equation $y = 3^x$ is graphed, which of the following values of x would produce a point closest to the x-axis?

a. $\frac{1}{3}$
b. $\frac{1}{6}$
c. $\frac{1}{9}$
d. $\frac{1}{12}$

46. $\mathbf{\log_8 42} =$

a. $(\log_{10} 8)(\log_{10} 42)$

b. $\frac{\log_{10} 42}{\log_{10} 8}$

c. $\log_{10} 8 - \log_{10} 42$

d. $\log_{10} 8 + \log_{10} 42$

47. What is the value of $\log_5 125$?

a. 3

b. 5

c. 10

d. 25

48. If $\log 2 \approx 0.402$ and $\log 5 \approx 0.581$, what is the approximate value of $\log 200$?

a. 0.037

b. 0.862

c. 1.443

d. 2.368

49. If x is a real number, for what values of x is the equation $\frac{4x+16}{4} = x + 4$ true?

a. impossible to determine

b. no values of x

c. some values of x

d. all values of x

50. On a recent test, Nikki wrote the equation $\frac{x^2-9}{x-3} = x + 3$. Which of the following statements is true about the equation she wrote?

a. The equation is never true.

b. The equation is always true.

c. The equation is always true, except when $x = 3$.

d. The equations is sometimes true when $x = 3$.

51. Given the equation $x^2 > x$, where x is a real number. Which of the following statements is true?

a. This equation is always true.

b. This equation is always true, except when $x = 0$ or $x = 1$.

c. This equation is sometimes true when $x = 0$ or $x = 1$.

d. This equation is never true.

52. Mackenzie wants to create several different 8-character passwords. She wants to use arrangements of the first 4 letters of her first name (mack), *followed by* arrangements of 4 digits of 1993, the year of her birth. How many different passwords can she create in this way?

a. 576

b. 512

c. 144

d. 128

53. A Christmas parade is made up of a convertible, 9 different floats, and a fire truck. If the convertible must be first, and the fire truck must be last, how many different ways can the parade be ordered?

a. 5040

b. 40,320

c. 362,880

d. 408,240

54. Britney and Christina were among 12 people selected to be a finalist for a radio contest. Two people from the group will be selected at random to win the contest. What is the probability that Britney and Christina will be the 2 people selected?

a. $\frac{1}{12}$

b. $\frac{1}{24}$

c. $\frac{1}{48}$

d. $\frac{1}{66}$

55. $(4y-2)^3 =$

a. $16y^3 - 72y^2 - 16y - 8$

b. $16y^3 + 72y^2 - 16y - 8$

c. $16y^3 - 72y^2 + 16y + 8$

d. $16y^3 + 72y^2 + 16y - 8$

56. How many terms does the binomial expansion of $(x^3 + 3y^2)^{18}$ contain?

a. 18

b. 19

c. 20

d. 23

57. What is the expanded form of $(x + y)^5$?

a. $x^5 + 4x^4y + 6x^3y^2 + 6x^2y^3 + 4xy^4 + y^5$

b. $x^5 + 5x^4y + 10x^3y^2 + 10x^2y^3 + 10xy^4 + y^5$

c. $x^5 + 5x^4y + 6x^3y^2 + 6x^2y^3 + 5xy^4 + y^5$

d. $x^5 + 4x^4y + 10x^3y^2 + 10x^2y^3 + 4xy^4 + y^5$

58. What is the 11th term in the arithmetic series below?

3, 7, 11...

a. 4

b. 39

c. 43

d. 51

59. What is the common ratio for the geometric sequence below?

12,36,108,324...

a. 2

b. 3

c. 4

d. 6

60. Given that $f(x) = 2x^2 + 1$ and $g(x) = 3x - 4$, what is $g(f(3))$?

a. 24

b. 17

c. 61

d. 53

61. Which expression represents $f(g(x))$ if $f(x) = x^2 + 2$ and $g(x) = (x - 4)$?

a. $x^2 - 8x + 18$

b. $x^2 - 2$

c. $x^2 + x + 2$

d. $x^2 - 4x + 18$

62. If $f(x) = x^2 + x + 2$ and $g(x) = 2(x - 1)^2$, which is an equivalent form of $f(x) + g(x)$?

a. $3x^2 + 2x + 2$

b. $x^2 + 3x + 4$

c. $6x^2 - x + 12$

d. $3x^2 - 3x + 4$

63. A science teacher is randomly distributing 12 beakers with milliliter labels and 4 beakers without beaker labels. What is the probability that the first beaker she hands out will have milliliter labels and the second beaker will *not* have labels?

a. $\frac{1}{5}$

b. $\frac{1}{4}$

c. $\frac{1}{6}$

d. $\frac{2}{7}$

64. The probabilities that James will try out for various sports and team positions are shown in the chart below.

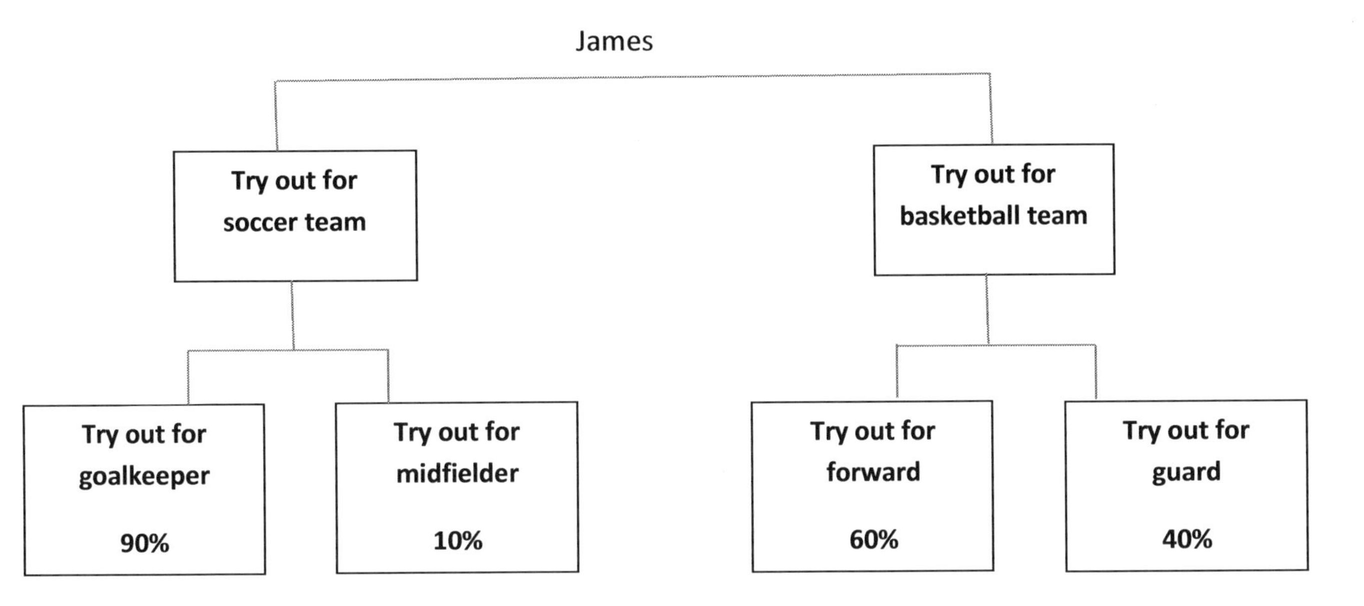

James will definitely try out for either soccer or basketball, but not both. The probability that James will try out for soccer and try out for goalkeeper is 63%. What is the probability that James will try out for basketball?

a. 70%
b. 60%
c. 30%
d. 20%

65. A temporary worker works on days as the need arises for their skills. The following list shows the number of days the worker was hired for a 5-month period.

$$6,11,15,8,14$$

If the mean of these data is approximately 11, what is the population standard deviation for these data? (Round the answer to the nearest tenth.)

a. 5.5

b. 10.8

c. 3.4

d. 6.8

Algebra 2-Questions: Section 2

1. What is the complete solution to the equation $|3 - 2x| = 11$?

a. $x = -4; x = 7$

b. $x = 4; x = 7$

c. $x = 4; x = -7$

d. $x = -4; x = -7$

2. What are the possible values of x in $|13 - 2x| = 5$?

a. $9 < x < -3$

b. $x = 3$ or $x = -9$

c. $-9 > x > -3$

d. $x = 3$ or $x = 9$

3. For a brunch, Shane bought several dozen cinnamon rolls and several dozen donuts. The cinnamon rolls cost $10 per dozen, and the donuts cost $7 per dozen. Shane bought a total of 20 dozen pastries and paid a total of $161. How many cinnamon rolls did he buy?

a. 6 dozen

b. 7 dozen

c. 8 dozen

d. 9 dozen

4. What is the solution to the system of equations shown below?

$$\begin{cases} x + 2y - z = 1 \\ 2x - y + 2z = 9 \\ x - 2y - 3z = -9 \end{cases}$$

a. No solution

b. Infinitely many solutions

c. $(\frac{1}{9}, 1, -\frac{1}{3})$

d. $(2,1,3)$

5. Teddy bears were $15 each and sock monkeys cost only $12 each. If Gillian bought 30 toys for her shop and paid $396, how many teddy bears did she buy?

a. 18

b. 16

c. 14

d. 12

6. What system of inequalities *best* represents the graph shown below?

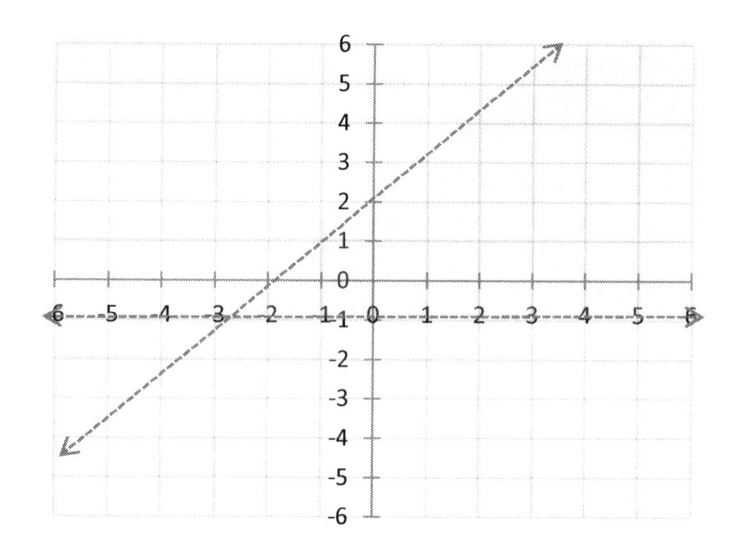

a. $y > -1$ and $y < x + 2$
b. $y > -1$ and $y > x + 2$
c. $y < -1$ and $y < x + 2$
d. $y > -1$ and $y > x + 2$

7. Which point lies in the solution set for the system $\begin{cases} y - 2x < 5 \\ 4y + 3x \geq 10 \end{cases}$?

a. (0,-3)

b. (3,1)

c. (1,1)

d. (-3,5)

8. Which system of linear inequalities is represented by this graph?

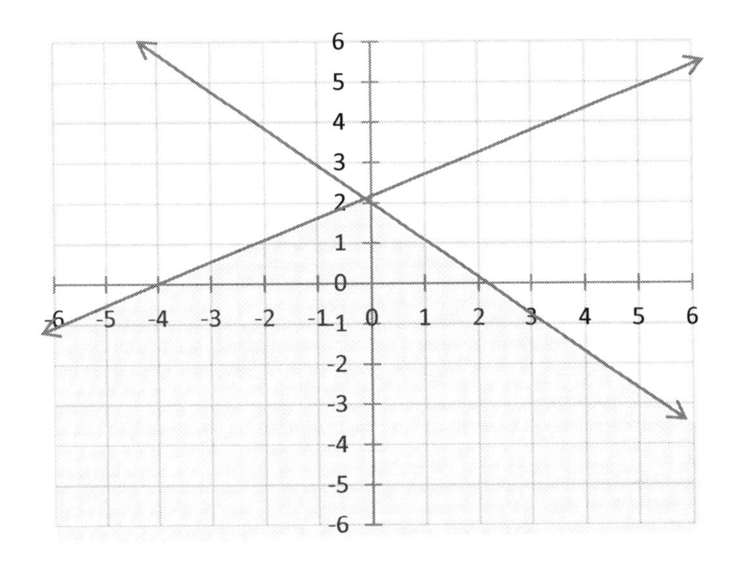

a. $\begin{cases} y > \frac{1}{2} - 2 \\ y \geq x + 2 \end{cases}$

b. $\begin{cases} y \leq \frac{1}{2}x + 2 \\ y \leq x - 2 \end{cases}$

c. $\begin{cases} y \geq -\frac{1}{2}x + 2 \\ y \leq -x + 2 \end{cases}$

d. $\begin{cases} y \leq \frac{1}{2}x + 2 \\ y \geq -x + 2 \end{cases}$

9. Which polynomial represents $(3x^2 - 4x + 3)(2x + 3)$?

a. $6x^3 - x^2 - 6 - 9$

b. $6x^3 + x^2 - 6 - 9$

c. $6x^3 + x^2 + 6 - 9$

d. $6x^3 + x^2 - 6 + 9$

10. $x - 3\overline{)4x^3 - 2x^2 + 4x + 2}$

a. $4x^2 - 10x + 34 + \frac{104}{x-3}$

b. $4x^2 - 10x - 34 - \frac{104}{x-3}$

c. $4x^2 + 10x + 34 + \frac{104}{x-3}$

d. $4x^2 + 10x - 34 + \frac{104}{x-3}$

11. $(-4x^2 + 7x + 3) - 2(2x^2 - 3x + 2)$

a. $13x - 1$

b. $x - 1$

c. $-8x^2 + x - 1$

d. $-8x^2 + 13x - 1$

12. Which expression is equivalent to $(5y^2 + 3)(5y - 3)$?

a. $25y^2 - 9$

b. $25y^3 - 9$

c. $25y^3 - 15y^2 + 15y - 9$

d. $25y^2 - 15y^2 + 15y - 9$

13. What is the volume of the figure below?

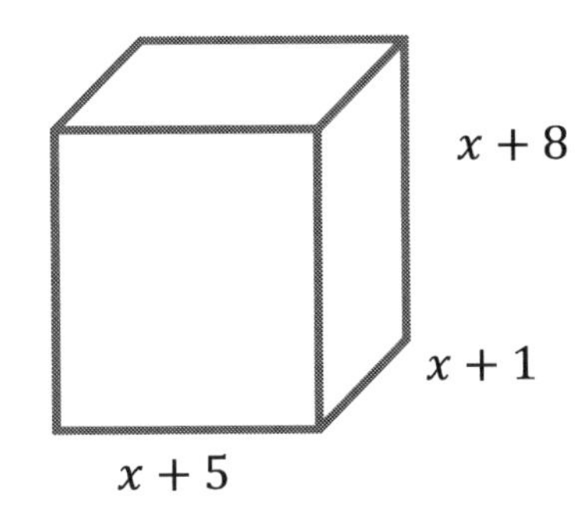

a. $x^3 + 13x^2 + 53x + 40$
b. $x^3 + 14x^2 + 53x + 40$
c. $x^3 + 13x^2 + 43x + 40$
d. $x^3 + 14x^2 + 43x + 40$

14. $27a^3 + c^3 =$

a. $(3a + c)(9a^2 - 3ac + c^2)$

b. $(3a - c)(9a^2 + 3ac + c^2)$

c. $(3a - c)(9a^2 + 6ac + c^2)$

d. $(3a + c)(3a + c)(3a + c)$

15. The total area of a rectangle is $9x^4 - 36y^2$. Which factors could represent the length times width?

a. $(3x^2 + 6y)(3x^2 + 6y)$

b. $(3x^2 - 6y)(3x^2 + 6y)$

c. $(3x - 6y)(3x - 6y)$

d. $(3x - 6y)(3x + 6y)$

16. Which expression shows the complete factorization of $18x^2 - 50$?

a. $(6x - 25)(3x + 2)$

b. $(9x + 5)(2x - 5)$

c. $6(x + 5)(x - 5)$

d. $2(3x + 5)(3x - 5)$

17. $\frac{x+2}{x+3} + \frac{6}{x^2-9} =$

a. $\frac{x^2+x+6}{x^2-9}$

b. $\frac{5x-6}{x^2-9}$

c. $\frac{x^2+5x+6}{x^2-9}$

d. $\frac{x^2-x-6}{x^2-9}$

18. Which product is equivalent to $\frac{3x^2-27}{3-x}$?

a. $-3(x + 3)$

b. $-3(x - 3)$

c. $3(x + 3)$

d. $3(x - 3)$

19. $\frac{x^2+3x}{x+7} \cdot \frac{x^2+5x-14}{x^2+x-6} =$

a. $\frac{x-2}{x+7}$

b. $\frac{x}{x+7}$

c. x

d. $x+2$

20. If $i = \sqrt{-1}$, which point shows the location of $2+3i$ on the plane?

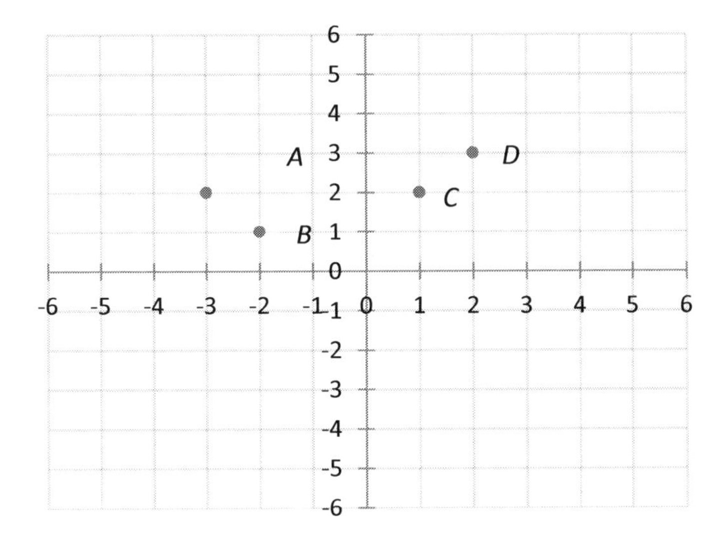

a. Point A

b. Point B

c. Point C

d. Point D

21. If $i = \sqrt{-1}$, what is the value of i^2?

a. 1

b. -1

c. i

d. $-i$

22. Which of the following complex numbers is represented by the point on the graph below?

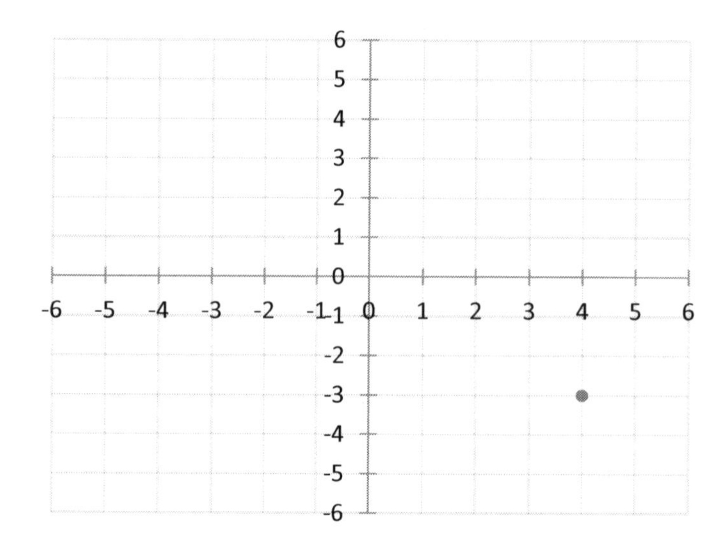

a. $3 - 4i$
b. $3 + 4i$
c. $4 - 3i$
d. $4 + 3i$

23. If $i = \sqrt{-1}$, then $2i(3i) =$

a.-6

b. 6

c.-12

d. 12

24. What is an equivalent form of $\frac{6}{1+i}$?

a.$\frac{6-i}{2}$

b. $6 - i$

c. $3 - i$

d. $\frac{3-i}{2}$

25. What is the product of the complex numbers $(7 + i)$ and $(7 - i)$?

a. 50

b. 48

c. $49 - i$

d. $14 - 7i$

26. What are to the solutions to the equation $5x^2 - 2x + 3 = 0$?

a. $x = \frac{1}{5} + \frac{14}{5}i; x = \frac{1}{5} - \frac{14}{5}i$

b. $x = \frac{1}{5} + \frac{\sqrt{14}}{5}i; x = \frac{1}{5} - \frac{\sqrt{14}}{5}i$

c. $x = \frac{1}{5} + 5i; x = \frac{1}{5} - 5i$

d. $x = \frac{1}{5}; x = \frac{14}{5}$

27. Jodi is solving the equation $x^2 - 6x = 8$ by completing the square. What number should be added to both sides of the equation to complete the square?

a. 9

b. 6

c. 12

d. 3

28. What are the solutions to the equation $3 + \frac{5}{x^2} = \frac{2}{x}$?

a. $x = \frac{1}{3} + \frac{\sqrt{14}}{3}i; x = \frac{1}{3} - \frac{\sqrt{14}}{3}i$

b. $x = -\frac{1}{3} + \frac{\sqrt{14}}{3}i; x = -\frac{1}{3} - \frac{\sqrt{14}}{3}i$

c. $x = \frac{1}{3}; x = \frac{14}{3}$

d. $x = \frac{1}{3} + \frac{\sqrt{14}}{3}; x = \frac{1}{3} - \frac{\sqrt{14}}{3}$

29. Which of the following sentences is true about the graphs of $y = -3(x+2)^2 + 1$ and $y = 3(x + 2)^2 + 1$?

a. The graphs have different shapes with different vertices.
b. The graphs have the same shape with different vertices.
c. One graph has a vertex that is a maximum, while the other graph has a vertex that is a minimum.
d. Their vertices are maximums.

30. Which of the following *most* accurately describes the translation of the graph $y = (x-3)^2 + 4$ to the graph of $y = (x+2)^2 - 2$?

a. up 5 and 3 to the right
b. down 6 and 5 to the left
c. up 1 and 4 to the left
d. down 6 and 2 to the right

31. What are the x-intercepts of the graph of $y = 6x^2 + 4x - 2$?

a. $\frac{1}{4}$ and $-\frac{2}{3}$

b. $-\frac{1}{4}$ and $\frac{2}{3}$

c. $-\frac{3}{4}$ and 2

d. $\frac{3}{4}$ and -2

32. Which is the graph of $y = (x+1)^2 - 3$?

a.

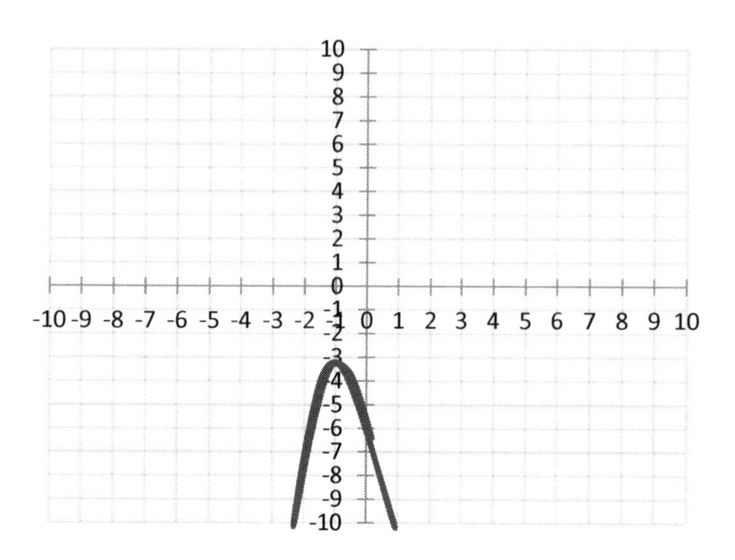

b.

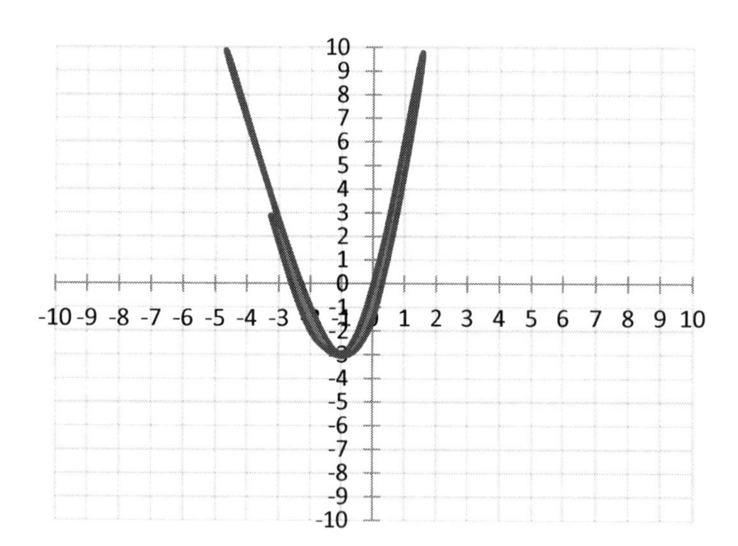

c.

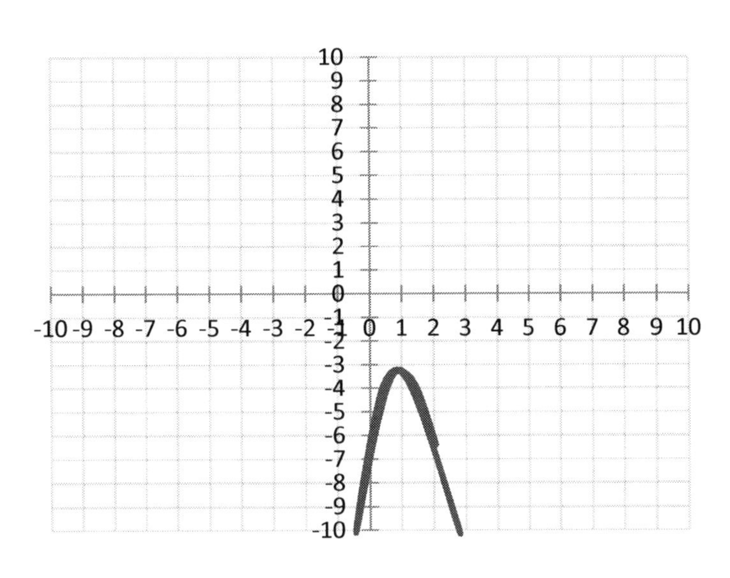
10
9
8
7
6
5
4
3
2
1
0
-1
-2
-3
-4
-5
-6
-7
-8
-9
-10
-10 -9 -8 -7 -6 -5 -4 -3 -2 -1 0 1 2 3 4 5 6 7 8 9 10

d.

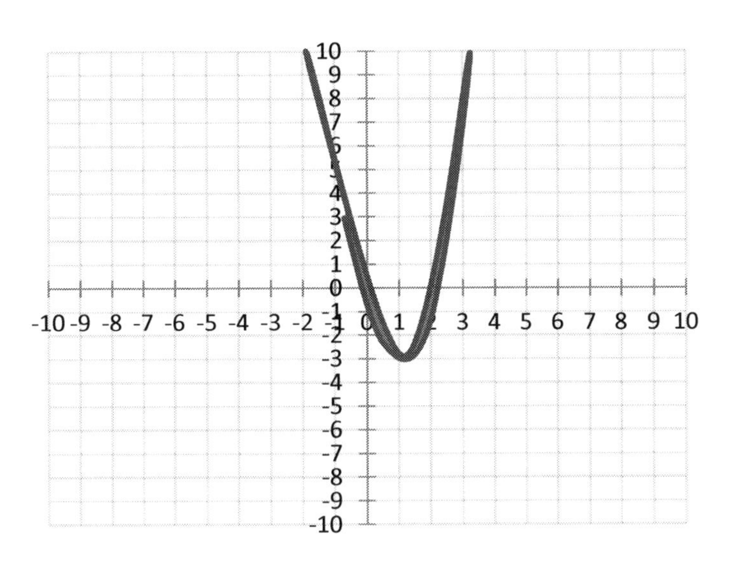
10
9
8
7
6
5
4
3
2
1
0
-1
-2
-3
-4
-5
-6
-7
-8
-9
-10
-10 -9 -8 -7 -6 -5 -4 -3 -2 -1 0 1 2 3 4 5 6 7 8 9 10

33. Which ordered pair is the vertex of $f(x) = x^2 - 8x + 19$?

a. $(-1, -5)$

b. $(-2, 0)$

c. $(8,1)$

d. $(4,3)$

34. The graph of $\left(\frac{x}{1}\right)^2 - \left(\frac{y}{4}\right)^2 = 1$ is a hyperbola. Which set of equations of represents the asymptotes of the hyperbola's graph?

a. $y = \frac{1}{4}x; y = -\frac{1}{4}x$

b. $y = \frac{1}{2}x; y = -\frac{1}{2}x$

c. $y = 4x; y = -4x$

d. $y = x; y = -x$

35. $\frac{(x-h)^2}{b^2} + \frac{(y-k)^2}{a^2} = 1$

Which standard form of a conic is represented by the equation above?

a. parabola
b. hyperbola
c. circle
d. volume

36. $2x^2 - 4y^2 + 8x - 32y + 12 = 0$

What is the standard form of the equation of the conic given above?

a. $\frac{(x-3)^2}{6} - \frac{(y+3)^2}{8} = 1$

b. $\frac{(y-1)^2}{2} - \frac{(x-2)^2}{3} = 1$

c. $\frac{(x+2)^2}{4} - \frac{(y-4)^2}{2} = 1$

d. $\frac{(y+5)^2}{3} - \frac{(x-6)^2}{4} = 1$

37. What is the solution to the equation $7^x = 19$?

a. $x = 12$

b. $x = \log_{10} 12$

$\log_7 19 = x$

c. $x = \log_{10} 7 + \log_{10} 19$

d. $x = \frac{\log_{10} 19}{\log_{10} 7}$

38. If $\log_{10} x = -1$, what is the value of x?

a. $\frac{1}{100}$

$10^{-1} = x$

b. -10

$\frac{1}{10}$

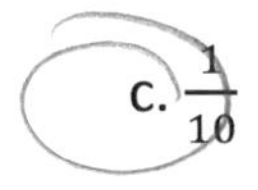

c. $\frac{1}{10}$

d. $\sqrt{\frac{1}{10}}$

39. Which equation is equivalent to $\log_4 \frac{1}{8} = x$?

a. $\left(\frac{1}{8}\right)^4 = x$

b. $4^x = \frac{1}{8}$

c. $\frac{1^4}{8} = x^4$

d. $4^{\frac{1}{8}} = x$

40. Which is the first *incorrect* step in simplifying $\log_3 \frac{3}{81}$?

Step 1: $\log_3 \frac{3}{81} = \log_3 3 + \log_3 81$

Step 2: $= 1 + 4$

Step 3: $= 5$

a. Step 1
b. Step 2
c. Step 3
d. Each step is correct.

41. Markus, Brian, Shane, and Gillian each worked the same math problem at the chalkboard. Each student's work is shown below. Their teacher said that while two of them had the correct answer, only one of them had arrived at the correct conclusion using correct steps.

Markus' Work

$x^2x^{-5} = \frac{x^2}{x^5}$

$= \frac{1}{x^3}; x \neq 0$

Shane's Work

$x^2x^{-5} = \frac{x^2}{x^5}$

$= x^3; x \neq 0$

Brian's Work

$x^2x^{-5} = \frac{x^2}{x^{-5}}$

$= \frac{1}{x^7}; x \neq 0$

Gillian's Work

$x^2x^{-5} = \frac{x^2}{x^{-5}}$

$= x^{-3}; x \neq 0$

a. Markus

b. Brian

c. Shane

d. Gillian

42. A student showed the following steps in his solution of the equation below, but his answer was not correct.

$$\log_{10}(x+3) + \log_{10}(x-3) = \log_3 7$$

Step 1: $\log_{10}(\frac{x+3}{x-3}) = 7$

Step 2: $\log_{10}(x+3) = -7$

Step 3: $x = -3 + -7$

Step 4: $x = -10$

In which step did he make his first error?

a. Step 1

b. Step 2

c. Step 3

d. Step 4

43. A limited edition model toy car will be sold in a quantity over time according to the equation $y = A\left(\frac{1}{3}\right)^{\frac{120}{60}}$, where A= the amount of toy cars available at the start of the sale and t =time in days. If 18,000 was the initial amount of cars available, how many cars will remain in 120 days?

a. 10000

b. 9000

c. 6000

d. 2000

44. Bacteria in a culture are growing exponentially with time, as shown in the table below.

Bacteria Growth

Day	Bacteria
0	300
1	600
2	900

Which of the following equations expressing the number of bacteria, y, present at any time, t?

a. $y = (900) \cdot (3)^t$
b. $y = 3^t$
c. $y = 3^t + 300$
d. $y = (300) \cdot (3)^t$

45. If the equation $y = 4^x$ is graphed, which of the following values of x would produce a point closest to the x-axis?

a. $\frac{1}{20}$
b. $\frac{1}{2}$
c. $\frac{1}{16}$
d. $\frac{1}{4}$

46. $\log_4 46 =$

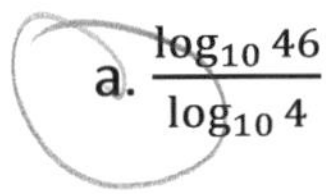

a. $\frac{\log_{10} 46}{\log_{10} 4}$

b. $\log_{10} 4 + \log_{10} 46$

c. $\log_{10} 4 - \log_{10} 46$

d. $(\log_{10} 4)(\log_{10} 46)$

47. What is the value of $\log_4 16$?

a. 4

b. 8

c. 2

d. 12

48. If $\log 2 \approx 0.178$ and $\log 7 \approx 0.385$, what is the approximate value of $\log 196$?

a. 0.776

b. 1.126

c. 1.304

d. 2.179

49. If x is a real number, for what values of x is the equation $\frac{x^2}{4} = x$ true?

a. impossible to determine

b. no values of x

c. some values of x

d. all values of x

50. On a recent quiz, Will wrote the equation $\frac{x^2+25}{x+5} = x - 5$. Which of the following statements is correct about the equation he wrote?

a. The equation is always true.

b. The equation is always true, except when $x = 5$.

c. The equation is sometimes true when $x = 5$.

d. The equation is never true.

51. Given the equation $y = \sqrt{x^2}$, where x is a real number. Which statement is valid for real values of y?

a. y is equal to negative real numbers

b. y is equal to non-negative real numbers

c. y cannot be defined

d. y is equal to 1

52. Albert wants to create several different 6-character screen names. He wants to use arrangements of the first 2 letters of his first name (al), *followed by* arrangements of 4 digits of 1987, the year of his birth. How many different screen names can he create in this way?

a. 24

b. 48

c. 64

d. 120

53. A music festival is made up of a famous solo artist, 6 different bands, and a famous vocal duo. If the solo artist must be first, and the vocal duo must be last, how many different ways can the music festival be ordered?

a. 120

b. 720

c. 1020

d. 5040

54. Christopher, Justin, and Ryan are among 15 students with who applied for a scholarship at a university. Three students from the group will be selected at random to receive the scholarship. What is the probability that Christopher, Justin, and Ryan will be the 3 students selected?

a. $\frac{1}{910}$

b. $\frac{1}{300}$

c. $\frac{1}{45}$

d. $\frac{1}{30}$

55. $(2y - 3)^4 =$

a. $16y^4 + 96y^3 + 216y^2 + 216y + 81$

b. $16y^4 - 96y^3 - 216y^2 + 216y + 81$

c. $16y^4 - 96y^3 + 216y^2 - 216y + 81$

d. $16y^4 + 96y^3 - 216y^2 + 216y - 81$

56. How many terms does the binomial expansion of $(2x^4 + y^3)^{30}$ contain?

a. 7

b. 29

c. 30

d. 31

57. What is the expanded form of $(x + y)^4$?

a. $x^4 + 4x^3y + 8x^2y^2 + 4xy^3 + y^4$

b.$x^4 + 5x^3y + 10x^2y^2 + 5xy^3 + y^4$

c. $x^4 + 4x^3y + 6x^2y^2 + 4xy^3 + y^4$

d. $x^4 + 4x^3y + 5x^2y^2 + 4xy^3 + y^4$

58. Paul was given the following mathematical induction to prove true, $3^n - 1$ is a multiple of 2. He needs to prove it is true. His work is shown below:

Step 1: Show it is true for $n = 1$

$3^1 - 1$

$3 - 1 = 2$

The statement is true.

Step 2: Show that if $n = k$ is true then $n = k + 1$ is also true

3^{k+1} is also $3 \cdot 3^k$ and is then split into $3 \cdot$ into $2 \cdot$ and $1 \cdot$ these are multiples of 2

$2 \cdot 3^k$ is a multiple of 2 therefore $3^k - 1$ is true (based upon the assumption made)

Is the work Paul did on this mathematical induction correct?

a. Yes, all steps are correct.
b. No, Step 1 is done incorrectly.
c. No, Step 2 is done incorrectly.
d. No, all steps are done incorrectly.

59. What is the 14th term in the arithmetic series below?

4,10,16,22...

a. 14
b. 46
c. 70
d. 88

60. What is the common ratio for the geometric sequence below?

$$2,10,50,250 \ldots$$

a. 2

b. 4

c. 5

d. 10

61. Given that $f(x) = 4x^2 + 5$ and $g(x) = 2x - 3$, what is $g(f(1))$?

a. 15

b. 19

c. 21

d. 23

62. Which expression represents $f\big(g(x)\big)$ if $f(x) = x^2 - 3$ and $g(x) = (x + 2)$?

a. $x^2 + 4x + 1$

b. $x^2 + 4$

c. $x^2 - x + 1$

d. $x^2 + 2x + 7$

63. If $f(x) = x^2 + 5x + 6$ and $g(x) = 4(x - 2)^2$, which is an equivalent form of $f(x) + g(x)$?

a. $5x^2 - 11x + 22$

b. $x^2 + 3x + 2$

c. $5x^2 + 21x + 10$

d. $4x^2 - 4x + 10$

64. On a certain day the chance of snow is 20% in Nashville and 70% in Buffalo. Assume that the chance of snow in the two cities is independent. What is the probability that it will *not* snow in either country?

a. 12%

b. 24%

c. 36%

d. 48%

65. A box contains 7 large blue circles, 9 large green circles, 3 small red circles, and 1 small blue circle. If a circle is drawn at random, what is the probability that it is blue, given that it is one of the large circles?

a. $\frac{8}{20}$

b. $\frac{7}{16}$

c. $\frac{3}{5}$

d. $\frac{1}{6}$

66. Agnes found the mean and standard deviation of the set of numbers below. If she adds 4 to each number, which of the following will result?

4,1,7,8,2,3

a. The mean will lower by 4.
b. The standard deviation will increase by 4 as the mean does not change.
c. The standard deviation will lower.
d. The standard deviation will not change as the mean will increase.

Algebra 2-Solutions: Section 1

1.

b. $\boldsymbol{x = -1; x = 2}$

The equation can be solved one of two ways as: $4 - 8x = 12$ or - $(4 - 8x) = 12$

To solve the first equation: $4 - 8x = 12$, we must subtract 4 from each side, leaving us with: $-8x = 8$.

Next we divide -8 from each side to get $x$alone: $x = -1$

We now know that one equation has $x = -1$, now we must solve the other equation: $-(4 - 8x) = 12$

We must distribute the – to the parts in the parenthesis: $-4 + 8x = 12$

Next we must add 4 to each side, to give us: $8x = 16$

To get x alone, we divide 8 from each side: $x = 2$

The complete solution for the problem is $x = -1; x = 2$

2.

c. $\boldsymbol{x = 3}$ **or** $\boldsymbol{x = 7}$

The absolute value equation can be solved in one of two ways as: $15 - 3x = 6$ or $-(15 - 3x) = 6$

To solve the first equation: $15 - 3x = 6$, we must subtract 15 from each side, leaving us with: $-3x = -9$.

Next we divide -3 from each side to get $x$alone: $x = 3$

We now know that one equation has $x = 3$, now we must solve the other equation: $-(15 - 3x) = 6$

We must distribute the – to the parts in the parenthesis: $-15 + 3x = 6$

Next we must add 15 to each side, to give us: $3x = 21$

To get x alone, we divide 3 from each side: $x = 7$

Therefore we know that the possible values of x can be $x = 3$or $x = 7$.

3.

a. 1875

To solve this problem we must first determine the percentage of people who bought tickets regular sale of tickets and then create an equation to solve for the total number of people attending the concert.

If there was 20 percent of concert attenders that bought their tickets during the pre-sale and the total number of ticket purchasers is 100 percent we can determine that those who bought their tickets during the regular sale is 80 percent.

We could make an equation look something like this: $20t + 80t = n$ with t representing the ticket rate and n being the total number of people attending the concert.

Since we know that $80t = 1500$, we can solve for t to use in our equation.

Divide each side by 80, $t = 18.75$

Now we can solve for n in our equation: $20(18.75) + 80(18.75) = n$

$$375 + 1500 = n$$

$$n = 1875$$

The total number of people attending the concert is 1875.

4.

c. (-2,1,2)

$2x + 3y - z = -3$ (a)

$x + 2y = 0 \rightarrow x = -2y$ (b)

$x - 2y + z = -2$ (c)

Since we could get x to equal $2y$ in equation (b), we can substitute equation (b) into (a) and (c) to get:

$-y - z = -3$ (a, substituted)

$\underline{-4y + z = -2}$ (c, substituted)

$-5y = -5$

Since z canceled each other out, we can solve for y: $y = 1$

Now we can solve for z by placing it in equation (a, substituted):

$-1 - z = -3$

$-z = -2$

Divide by -1: $z = 2$

Solve for x by placing the solution for y in equation (b):

$x = -2(1)$

$x = -2$

5.

a. 6

We can do this problem algebraically by making p =the amount of 8 pairs of socks bought and s =the amount of 10 pairs of socks bought

We can create the problem: $8p + 10s = 132$ to show the total amount of money spent on the pastries

Also we can make another algebra equation for this word problem: $p + s = 15$ to show the total packages of socks bought. This equation can be substituted into the first to find how many 10 pairs of socks were bought. We can do this by substituting for p(8 pairs of socks). This equation would be: $p = 15 - s$

Now solve for s: $8(15 - s) + 10s = 132$

Distribute 8 to the parenthesis:$120 - 8s + 10s = 132$

Combine the variables together: $2s + 120 = 132$

Subtract 120 from each side: $2s = 12$

Divide 2 from each side to determine the value of s: $s = 6$

The store manager purchased 6 10 pairs of socks packages and 9 8 pairs of socks packages.

6.

d. $y < 3$ and $y > x - 1$

The two lines on the graph are best represented by $y < 3$ and $y > x - 1$. The line is drawn as a dashed line to show the points on the line do not satisfy the inequality. The region we wish to find is below the $y < 3$ line and to the left of the $y > x - 1$.

7.

b. (1,2)

Out of the four answer choices, only (1,2) are the only point that lie in the solution set for the system to make it true as shown:

$3(2) + 2(1) > -6$

$6 + 2 > -6$

$8 > -6$ (True)

$-3(2) + 1 \geq -6$

$-6 + 1 \geq -6$

$-5 \geq -6$ (True)

8.

b. $\begin{cases} y > \frac{1}{3}x - 4 \\ y < x - 3 \end{cases}$

The two lines on the graph are best represented by $y > \frac{1}{3}x - 4$ and $y < x - 3$. The line is drawn as a dashed line to show the points on the line do not satisfy the inequality.

9.

c. $2x^3 + 6x^2 - 10x - 8$

To solve this problem, each term of the expression is multiplied by every term of the other expression. Then like terms in the product are added together. You can use a vertical or horizontal format to find the product. Either format is acceptable to use. In this equation, we will use the horizontal format:

$(2x^2 - 2x - 2)(x + 4)$

$2x^3 + 8x^2 - 2x^2 - 8x - 2x - 8$

Combine like terms to get the final product: $2x^3 + 6x^2 - 10x - 8$

10.

a. $\mathbf{3x^2 + 10x + 45 + \frac{182}{x-4}}$

For each term we would take out the number that when multiplied would give us the term that we are dividing by as shown below:

$$\begin{array}{r} 3x^2 + 10x + 45 + \frac{182}{x-4} \\ x-4\overline{)3x^3 - 2x^2 + 5x + 2} \\ \underline{3x^3 - 12x^2} \qquad\qquad \\ 10x^2 + 5x \qquad \\ \underline{10x^2 - 40x} \qquad \\ 45x + 2 \\ \underline{45x - 180} \\ 182 \end{array}$$

Change signs and add

Change signs and add

Change signs and add

Therefore, $\frac{3x^3-2x^2+5x+2}{x-4} = 3x^2 + 10x + 45 + \frac{182}{x-4}$

11.

d. $\mathbf{-9x^2 + 11x - 1}$

To solve this equation you simply distribute the -3 to the equation then combine like terms as shown:

$(-3x^2 + 5x + 2) - 3(2x^2 - 2x + 1)$

$-3x^2 + 5x + 2 - 6x^2 + 6x - 3$

$-9x^2 + 11x - 1$

12.

c. $\mathbf{9y^3 + 3y^2 - 3y - 1}$

To solve this problem, you simply use the FOIL method as shown below:

$(3y^2 - 1)(3y + 1)$

$9y^3 + 3y^2 - 3y - 1$

Since there aren't any like terms the equation remains as $9y^3 + 3y^2 - 3y - 1$

13.

b. $\boldsymbol{x^3 + 10x^2 + 31x + 30}$

To find the volume we must remember the rectangular solid which is: Volume$= l \times w \times h$

Our equation would be: $(x + 3)(x + 2)(x + 5)$

Now we must multiply as follows: $x^2 + 2x + 3x + 6(x + 5)$

Combine like terms before multiplying $(x + 5)$: $x^2 + 5x + 6$

Now multiply $(x^2 + 5x + 6)$ with $(x + 5)$: $(x^2 + 5x + 6)(x + 5)$

$$x^3 + 5x^2 + 5x^2 + 25x + 6x + 30$$

Combine like terms to get our final answer: $x^3 + 10x^2 + 31x + 30$

14.

a. $\boldsymbol{(3a + c)(6a^2 - 3ac + c^2)}$

To find the equation to the answer we must work out the various possibilities. A simple way to do this is by checking the choices given and multiplying them out. With the equation $(3a + c)(6a^2 - 3ac + c^2)$ the work is shown below:

$18a^3 - 9a^2c + 3ac^2 + 6a^2c - 3ac^2 + c^3$

Then combine like terms to get your answer: $18a^3 - 3a^2c + c^3$

The ac^2 variable cancels itself out leaving the answer you see above.

15.

a. $\boldsymbol{(5x^2 - 4y)(5x^2 + 4y)}$

The factors that represent the total area of this rectangle must have the xy factor cancel each other out when using the FOIL method to solve the area. Only answers a and d do that, we must also consider that the total area includes a x^4. This can only be achieved by having x^2 times x^2 thus leading us to answer choice a.

This also can be factored to check to see if we are correct:

$(5x^2 - 4y)(5x^2 + 4y)$

$5x^4 + 20x^2y - 20x^2y - 16y^2$

The x^2y variable cancels itself out leaving the total area of $25x^4 - 16y^2$

16.

c. $(\boldsymbol{x - 1 + y})(\boldsymbol{x - 1 + y})$

To find the product of factors equivalent to $(x-1)^2 + y^2$, we must simply look at other ways it could possibly be written. $(x-1)^2$ could be separated as: $(x-1)(x-1)$.

We also need to do this with our y^2. This can be included into the $(x-1)(x-1)$ as $(x - 1 + y)(x - 1 + y)$

17.

d. $\boldsymbol{\frac{x^2-2x}{x^2-x-6}}$

To arrive at this answer, we must work through the rational expression:

$\frac{x-4}{x-3} + \frac{8}{x^2-x-6}$

$\boldsymbol{\frac{x-4}{x-3} + \frac{8}{(x-3)(x+2)}}$

$\boldsymbol{\frac{x-4(x+2)}{x-3(x+2)} + \frac{8}{(x-3)(x+2)}}$

$\boldsymbol{\frac{x^2+2x-4x-8}{x-3(x+2)} + \frac{8}{(x-3)(x+2)}}$

The bottom is the same so it can be multiplied once using the FOIL method and then we can combine like terms.

$\boldsymbol{\frac{x^2+2x-4x-8+8}{x^2-x-6}}$

The 8's cancel each other out leaving us with:

$\frac{x^2-2x}{x^2-x-6}$

18.

b. $\frac{4a^7}{b^7c^2}$

Our goal in this problem is to get the exponent all positive. First we need to solve $(a^{-2}b^3c)^3$ to simplify the exponents.

$$\frac{4ab^2c^{-1}}{a^{-6}b^9c^3}$$

Now we take the negative exponents and add them to the positive exponents leaving the exponent in a place where they remain positive, giving us our solution, as shown:

$$\frac{4a^7}{b^7c^2}$$

19.

a. $\frac{x+5}{x-5}$

To find the simplest form of this equation we first need to find the factors of the equation

$\frac{x^2+5x}{x^2+2x} \cdot \frac{x^2-4}{x^2-7x+10}$

$\frac{x(x+5)}{x(x+2)} \cdot \frac{(x+2)(x-2)}{(x-2)(x-5)}$

Now cancel out variables that are the same to receive our final answer: $\frac{x+5}{x-5}$

20.

b. Point B

The real part of the equation is –2 and the imaginary part of the equation is 1, which means that on the complex plane, the point on the graph is (–2, 1).

21.

a. i

Since $\sqrt{-1}$ times $\sqrt{-1}$ equals -1, then ii equals -1 and i^2 equals -1. We need to keep this in mind when simplifying expressions with more than one i.

$i^5 = (ii)(ii)i$

$i^5 = (-1)(-1)i$

$i^5 = (1)i$

$i^5 = i$

22.

c. $2 + 5i$

The real part of the equation is 2 and the imaginary part of the equation is 5, which means that on the complex plane, the point on the graph is (2, 5). Therefore the equation $2 + 5i$.

23.

b. -15

$3i(5i)$ can be simplified by multiplying the real numbers then the imaginary numbers. 3 multiplied by 5 is 15 and ii equals -1. 15 times -1 equals -15.

24.

d. $\frac{5-i}{13}$

To find the equivalent form of $\frac{2}{5+i}$ we need to multiply the complex number to the numerator to eliminate it from the denominator with the opposite of $5 + i$, $5 - i$. This is shown below:

$$\frac{2}{5+i} \cdot \frac{5-i}{5-i}$$

The numerator becomes $10 - i$, while the denominator becomes: $25 - i + i - i^2$.

The i cancel each other out leaving $25 - i^2$. i^2 is -1 and since there is a negative in front of it, the 1 will become positive giving us $25 + 1$.

Adding this will give us 26, the fraction is now: $\frac{10-i}{26}$

The numbers in the fraction can be reduced since both are divisible by 2, giving us the answer of: $\frac{5-i}{13}$

25.

c. 17

When finding the product of complex numbers, you are using the FOIL method to arrive at the answer as shown below:

$(4 + i)(4 - i)$

$16 + i - i - i^2$

The i cancel each other out, leaving: $16 - i^2$

i^2 equals -1 and we can place it into the equation: $16 - (-1)$ or $16 + 1 = 17$

26.

d. $\boldsymbol{x = \frac{1}{4} + \frac{\sqrt{23}}{4}i; x = \frac{1}{4} - \frac{\sqrt{23}}{4}i}$

The equation is written in standard form: $2x^2 - x + 3 = 0$

We divide every term by 2 so that the coefficient of x^2 will be 1: $x^2 - \frac{1}{2}x + \frac{3}{2} = 0$

Next we write the parentheses and move the constant term to the right hand side:

$(x^2 - \frac{1}{2}x \quad) = -\frac{3}{2}$

Now we multiply the coefficient of x by $\frac{1}{2}$ and square the product: $\left(-\frac{1}{2} \cdot \frac{1}{2}\right)^2 = \frac{1}{16}$

Then we add $\frac{1}{16}$ to both sides of the equation: $x^2 - \frac{1}{2}x + \frac{1}{16} = -\frac{3}{2} + \frac{1}{16}$

Now we simplify and solve for x: $\left(x - \frac{1}{4}\right)^2 = -\frac{23}{16}$ simplified

$x - \frac{1}{4} = \pm\sqrt{-\frac{23}{16}}$ square root of both sides

$x - \frac{1}{4} = \pm\frac{\sqrt{23}}{4}i$ solved

27.

d. 25

Standard form is $ax + by + c = 0$, where c is a constant. When you complete the square, you first move the constant to one side and the variables with coefficients in front of them to the other.

For this problem, that has already been done for us. The next step is to divide by 2 and square the resulting answer as shown: $-\frac{10}{2} = -5$

$(-5)^2 = 16$

Now add this number to both sides, therefore the answer is 25.

28.

b. $\boldsymbol{x = -\frac{1}{4} + \frac{\sqrt{31}}{4}i; x = -\frac{1}{4} - \frac{\sqrt{31}}{4}i}$

We first need to get the equation in standard form. This can be done by multiplying both sides by x^2 as shown:

$x^2(2 + \frac{4}{x^2} = -\frac{1}{x})$

$2x^2 + 4 = -x$

$2x^2 + x - 4 = 0$

Now that it is in standard form, we can now use the quadratic formula to solve:

$$x = \frac{-1 \pm \sqrt{(1)^2 - 4(2)(4)}}{2(2)} = \frac{-1 \pm \sqrt{-31}}{4}$$

$x = -\frac{1}{4} + \frac{\sqrt{31}}{4}i$ or $x = -\frac{1}{4} - \frac{\sqrt{31}}{4}i$

29.

d. The graphs have the same shape with different vertices.

The two equations we have are already in the best form to see what they look like.

The vertex form of a parabola (opening up or down) is:

$y = a(x - h)^2 + k$

The vertex of the parabola is (h, k).

If a is positive, the parabola opens upward, and the vertex is a minimum.

If a is negative, the parabola opens downward, and the vertex is a maximum.

Also, the bigger a is, the faster the parabola grows, and the skinnier it will look. The smaller a is, the slower the parabola grows.

We have two parabolas that are almost identical, but have different vertices:

$y = 2(x - 4)^2 + 3$ vertex $(-4, 3), a = 2$

$y = 2(x + 4)^2 + 3$ vertex $(4, 3), a = 2$

$a = 2$ for both, so both parabolas open upwards and their vertices are minimums. This makes choice b false.

Also since they have the same a, they both increase at the same rate, they have the same shape. This makes choice c false.

A is also clearly false since $a = 2$, for both; the only way choice a could be true would be if a was positive for one and negative for the other.

So the only correct answer is d.

30.

a. up 4 and 4 to the right

To move a function up, you add outside the function: $\mathrm{f(x)} + \mathrm{b}$ is $\mathrm{f(x)}$ moved up b units. Moving the function down works the same way; $\mathrm{f(x)} - \mathrm{b}$ is $\mathrm{f(x)}$ moved down b units.

In the first equation we see $y = (x + 2)^2 - 1$

b is -1

In the second equation $y = (x - 2)^2 + 3$ and b is 3

Therefore the b is moving from -1 to 3 on the graph meaning it will move up 4 places

To shift a function left, add inside the function's argument: $f(x + b)$ gives $f(x)$ shifted b units to the left. Shifting to the right works the same way; $f(x - b)$ is $f(x)$ shifted b units to the right.

In our first equation the b is 2 and in the second equation the b is -2.

The common mistake is to think that $f(x + 2)$ moves $f(x)$ to the right by two, because +2 is to the right. But the left-right shifting is backwards from what you might have expected. Adding moves you left; subtracting moves you right.

Therefore it is moving of the graph from -2 to 2 meaning it will move right 4 places.

31.

c. -1 and $\frac{1}{3}$

Let $y = 0$ to solve for x-intercepts:

$6x^2 + 4x - 2 = 0$

$(2x + 2)(3x - 1) = 0$

$2x + 2 = 0; 2x = -2; x = -1$

$3x - 1 = 0; 3x = 1; x = \frac{1}{3}$

The x-intercepts are -1 and $\frac{1}{3}$

32.

c.

Graph c is the best representation of $y = (x - 2)^2 + 4$. To better understand this, we need to take the equation from vertex form to standard form to get a clearer picture of how it would be graphed.

$(x - 2)(x - 2) + 4$

$x^2 - 2x - 2x + 4 + 4$

$y = x^2 - 4x + 8$

Our vertices would be $(2,4)$ and would be the point that we would see graphed on the parabola. We find the vertices by looking at our original equation the h and k would be our points.

33.

d. (5,0)

To solve for the vertices in a standard form, we must first find x. The equation to find x is $x = -\frac{b}{2a}$

Once we find x, it can be plugged into our equation to find y.

$x = -\frac{-10}{2(1)}$

$x = \frac{10}{2}$

$x = 5$

$y = (5)^2 - 10(5) + 25$

$y = 25 - 50 + 25$

$y = 0$

Our ordered pair for the vertex is $(5,0)$

34.

b. $\frac{(x-h)^2}{a^2} - \frac{(y-k)^2}{b^2} = 1$

Choice a is the standard equation of a parabola with vertex at the origin and opens to the right or left.

Choice c is the standard equation for the circle of radius r centered at the origin.

Choice d is the volume of a sphere. Therefore our answer is b.

35.

d. $y = \frac{2}{3}x; y = -\frac{2}{3}x$

The slopes of the two asymptotes will be of the form $m = \pm\frac{a}{b}$.

The number under x is a and the number under y is b.

Therefore our fraction would be$\pm\frac{2}{3}x$

36.

a. $\frac{(x-3)^2}{7} - \frac{(y+1)^2}{2} = 1$

We need to get the problem into standard form. First we need to put similar variables in the problem near to each other: $2x^2 - 12x - 7y^2 - 14y - 4 = 0$

Move 4 to the opposite side: $2x^2 - 12x - 7y^2 - 14y = 4$

Next we need to get the squared variables not to have a number in front:

$2(x^2 - 6x\quad) - 7(y^2 + 2y\quad) = 4$

We must put in the missing number from the square to place in their respected equations and add to the opposite side: $2(x^2 - 6x + 9) - 7(y^2 + 2y + 1) = 4 + 9 + 1$

Factor the double square and add like terms on the opposite side:

$$2(x-3)^2 - 7(y+1)^2 = 14$$

Divide 2 and 7 from each side to get 14 on the opposite side to be 1 which will make it into standard form and give us the answer of:

$$\frac{(x-3)^2}{7} - \frac{(y+1)^2}{2} = 1$$

37.

d. $\boldsymbol{x = \frac{\log_{10} 11}{\log_{10} 3}}$

The equation $3^x = 11$ is a logarithm. To solve for x, you must divide 3 from each side to get an answer. The answer for this equation is: $x = \frac{\log_{10} 11}{\log_{10} 3}$

38.

a. $\boldsymbol{\frac{1}{1000}}$

To find out the value of x, we must divide the $\log_{10}$ from each side and since it cannot be a negative it is written as followed: $\frac{1}{10^3}$

Now solve the fraction: $\frac{1}{(10)(10)(10)} = \frac{1}{1000}$

39.

c. $\boldsymbol{3^x = \frac{1}{6}}$

$\log_3 \frac{1}{6} = x$ can be written as the $\log$ raised to the x power equaling whatever number right next to the $\log$. In this case our $\log$ is 3 so it would be: $3^x = \frac{1}{6}$

40.

d. Each step is correct.

The equation $\log_5 \frac{5}{125}$ is solved correctly giving the answer of -2.

41.

b. Jordan

To solve x^4x^{-6} you must take the negative exponent and place it on the bottom to make it positive: $\frac{x^4}{x^6}$

Subtract 6-4 to make the numerator 1 and the denominator the remaining exponent: $\frac{1}{x^2}$

42.

c. Step 3

The first two steps were worked out correctly. It was not until Step 3 that the first error occurred. Instead of adding 14 to 20, it should have been subtracted. The problem is worked out correctly below:

Step 1: $\mathbf{\log_3(x+7)(2) = 20}$

Step 2: $\mathbf{\log_3(2x+14) = 20}$

Step 3: $\mathbf{2x = 6}$

Step 4: $x = 3$

43.

b. 1250

To solve this problem, simply plug in the numbers given into the equation: $y = 20000\left(\frac{1}{4}\right)^{\frac{400}{200}}$

$\frac{400}{200}$ can be reduced to 2 making the equation now: $y = 20000\left(\frac{1}{4}\right)^2$

$\frac{1}{4} \times \frac{1}{4} = \frac{1}{16}$

$y = 20000\left(\frac{1}{16}\right)$ or $y = \frac{20000}{16}$

$y = 1250$

The car will depreciate in value of $1250 in the course of 400 days.

44.

a. $\mathbf{y = (50) \cdot (2)^t}$

To create an equation for this word problem, we must remember that we need to use a variable that remains as a starting point. The mold begins at 50 mold spores and increases by each day.

From the sample chart we are given, we can see that the number doubles with each day. Therefore we can make the conclusion that our starting point of 50 is multiplied by 2 that has the exponent of t.

45.

c. $\mathbf{\frac{1}{9}}$

To solve for 3^x, we will allow $y = 0$.

$3^x = 0$

$x = \frac{1}{3^2}$

$x = \frac{1}{9}$

Therefore our x-axis would be $\frac{1}{9}$

46.

b. $\mathbf{\frac{\log_{10} 42}{\log_{10} 8}}$

Another way that this logarithm can be written out is by dividing the number that is next to the logarithm by the log base making the log base to be now a $\log_{10}$, with the problem being $\frac{\log_{10} 42}{\log_{10} 8}$

47.

a. 3

The value of $\log_5 125$ can be determined by knowing that 125 is 5^3. Therefore the value is 3.

48.

d. 2.368

To find the approximate value of $\log 200$, we first must determine the 2 and 5 factors of 200.

Factors of 200 using only 2 and 5 are: $2 \cdot 2 \cdot 2 \cdot 5 \cdot 5$

Next we add up the values of $\log 2$ and $\log 5$ to get our approximate value of $\log 200$

$$0.402 + 0.402 + 0.402 + 0.581 + 0.581 = 2.368$$

The approximate value of $\log 200$ is 2.368

49.

d. all values of x

The equation $\frac{4x+16}{4} = x + 4$ is equal to each other since it can also be written as $\frac{4(x+4)}{4} = x + 4$

The 4's that are in the fraction would cancel one another out leaving you with $x + 4 = x + 4$

Both are the same problem thus all values of x would be true for this equation.

50.

c. The equation is always true, except when $x = 3$.

The equation is always true when x is equal to any number except 3.

For example when $x = 2$, the equation looks like this: $\frac{(2)^2-9}{2-3} = 2 + 3$

$$\frac{4-9}{-1} = 5$$

$$-\frac{5}{-1} = 5; 5 = 5$$

This equation is true.

Now we need to prove that when $x = 3$, the equation is false.

$$\frac{(3)^2 - 9}{3 - 3} = 3 + 3$$

$$\frac{9 - 9}{3 - 3} = 6$$

$$\frac{0}{0} = 6$$

Zero does not equal 6 therefore when $x = 3$, the equation is false.

51.

b. This equation is always true, except when $x = 0$ or $x = 1$.

The equation $x^2 > x$ is always true when it is any real number that is not 0 or 1. For example when $x = -1$, we have $(-1)^2 > -1; 1 > -1$

This is true.

When we have $x = 0$ or $x = 1$, the numbers are the same as shown below:

$(0)^2 > 0$ The zeros would be equal to each other, not greater than.

$(1)^2 > 1$ The ones would also be equal to each other, not greater than one another.

52.

a. 576

There are 24 different combinations of mack that Mackenzie could use ($4!$ which is 4 factorial or $4 \cdot 3 \cdot 2 \cdot 1 = 24$)

There are 24 different combinations of 1993 that Mackenzie could use (Again $4!$ which is 4 factorial or $4 \cdot 3 \cdot 2 \cdot 1 = 24$)

You multiply both parts $24 \cdot 24$ and get 576.

53.

c. 362,880

If the convertible always is first and the fire truck is always last, this means that the 9 different floats can be placed in any order. To find this number, we can use $9!$ (which is 9 factorial)

$$9 \cdot 8 \cdot 7 \cdot 6 \cdot 5 \cdot 4 \cdot 3 \cdot 2 \cdot 1 = 362{,}880$$

54.

d. $\frac{1}{66}$

The first finalist can be chosen in 12 ways
The second finalist can be chosen in 11 ways, after the first person is selected.

So there are $12 \cdot 11 = 132$ different choices for first and second choices.

In this matter, the order is not important. There are twice as many options as necessary; therefore we need to divide by 2.

There are a total of 132 groups of 2 finalists. One of those groups will contain Britney and Christina, therefore $\frac{1}{66}$ is the probability of both being selected.

55.

a. $16y^3 - 72y^2 - 16y - 8$

Below is the expanded binomial expression and the steps to arrive at our answer:

$(4y - 2)^3 = (4y - 2)(4y - 2)(4y - 2)$

$(4y^2 - 16y + 4)(4y - 2)$

$16y^3 - 8y^2 - 64y^2 - 32y + 16y - 8$

Combine like terms to get the answer: $16y^3 - 72y^2 - 16y - 8$

56.

b. 19

In a linear equation (a binomial to the first power) there are two terms. For a cubic (binomial cubed) there are 4 terms. There is a pattern that emerges from this: a binomial expansion of $(a + b)^n$ will have $n + 1$ terms.

In the problem given $(x^3 + 3y^2)^{18}$ our binomial expansion contains 19 terms.

57.

b. $x^5 + 5x^4y + 10x^3y^2 + 10x^2y^3 + 10xy^4 + y^5$

The simplest way to find the expanded binomial when we have a $(x + y)^n$ or a larger exponent is to use Pascal's triangle.

The equation is raised to the 5th power which means that there will be 6 terms. You would then look at the triangle to determine the number that will be in from of each expression. For this problem, you would look at the 6th row and find the numbers: 1,5,10,10,5,1

The exponents for each coefficient would go in decreasing order for x and increased order for y, giving us answer choice b.

58.

c. 43

To find the 11th number in the arithmetic series we must use the formula $a_n = a_1 + (n-1)d$

Our common difference is 4 making $d = 4$

The number we are wanting to find is the 11th term, this is n.

a_1 is the first number in the sequence which is 3

Our equation: $a_{11} = 3 + (11-1)4$

$3 + (10)4$

$3 + 40$

$a_{11} = 43$

59.

b. 3

To find the common ratio, we have to divide a pair of terms. It does not matter which pair is picked, as long as they are right next to each other:

$$\frac{36}{12} = 3; \frac{108}{36} = 3; \frac{324}{108} = 3$$

Each number that is right next to each other has the same common ratio of 3.

60.

d. 53

To solve this problem you must place $f(x)$ equation into the x part of the $g(x)$ equation and the x in the $f(x)$ equation equals 3.

This is shown below:

$3(2(3)^2 + 1 - 4$

$3(2(9) + 1) - 4$

$3(18 + 1) - 4$

$3(19) - 4$

$57 - 4 = 53$

61.

a. $x^2 - 8x + 18$

We need to replace the x in $f(x)$ with the entire expression of $g(x)$ as shown below:

$$(x-4)^2 + 2$$

Now we will factor and combine like terms to get our answer:

$(x-4)(x-4) + 2$

$x^2 - 4x - 4x + 16 + 2$

$x^2 - 8x + 18$

62.

d. $3x^2 - 3x + 4$

To find the sum of $f(x) + g(x)$, first we must factor out the equation $g(x)$ before we can combine like terms.

$2(x-1)^2$

$2(x-1)(x-1)$

$2(x^2 - 2x + 1)$

Multiply $2x^2 - 4x + 2$

Add $f(x)$ and $g(x)$: $2x^2 - 4x + 2 + x^2 + x + 2$

$3x^2 - 3x + 4$

63.

a. $\frac{1}{5}$

There are 16 total beakers. For the first student to receive a beaker with milliliter labels is $\frac{12}{16}$, for the second student to receive a beaker without milliliter labels is $\frac{4}{15}$. To find out the probability of the first two students receiving the beakers in this order, we would multiply $\frac{12}{16}$ by $\frac{4}{15}$

$\frac{12}{16} \cdot \frac{4}{15} = \frac{48}{240}$

Simplify: $\frac{48}{240} = \frac{1}{5}$

There is $\frac{1}{5}$ probability that the beakers will be handed out in this order.

64.

c. 30%

We need to look at the problem in this way: The chance that he's trying out for soccer and goalkeeper is 63%, meaning that two numbers were multiplied to get 63%

Since we know one of the numbers, we can find the other

$0.9(\text{or } 90\%)x = 0.63(\text{or } 63\%)$

Divide 63 by 9 which is 7

Therefore the probability that James will try out for soccer is 0.70 (70%)

Meaning the probability that he'll try out for basketball is $100\% - 70\% = 30\%$

65.

c. 3.4

To find the population standard deviation we must find the variance. To find this we must take the difference of each number from the mean. We will take each difference, square it, and then average the result:

$$\frac{(-5)^2 + 0^2 + 4^2 + (-3)^2 + 3^2}{5}$$

$$\frac{25 + 0 + 16 + 9 + 9}{5} = \frac{59}{5}$$

Our variance is $\frac{59}{5}$ or 11.8

Next we take the square root of the variance to get the population standard deviation: $\sqrt{11.8} = 3.4351128046534$

We will round this number to the nearest tenth to give us a population standard deviation of 3.4

Question Number	Correct Answer	Standard
1	B	1.0
2	C	1.0
3	A	2.0
4	C	2.0
5	A	2.0
6	D	2.0
7	B	2.0
8	B	2.0
9	C	3.0
10	A	3.0
11	D	3.0
12	C	3.0
13	B	3.0
14	A	4.0
15	A	4.0
16	C	4.0
17	D	7.0
18	B	7.0
19	A	7.0
20	B	5.0
21	A	5.0
22	C	5.0
23	B	6.0
24	D	6.0
25	C	6.0
26	D	8.0
27	D	8.0
28	B	8.0
29	D	9.0
30	A	9.0
31	C	10.0
32	C	10.0
33	D	10.0
34	B	16.0
35	D	16.0
36	A	17.0
37	D	11.1
38	A	11.1
39	C	11.1

40	D	11.2
41	B	11.2
42	C	11.2
43	B	12.0
44	A	12.0
45	C	12.0
46	B	13.0
47	A	14.0
48	D	14.0
49	D	15.0
50	C	15.0
51	B	15.0
52	A	18.0
53	C	18.0
54	D	19.0
55	A	20.0
56	B	20.0
57	B	20.0
58	C	22.0
59	B	22.0
60	D	24.0
61	A	24.0
62	D	25.0
63	A	PS1.0
64	C	PS2.0
65	C	PS7.0

Algebra 2-Solutions: Section 2

1.

a. $\boldsymbol{x = -4; x = 7}$

The equation can be solved one of two ways as: $3 - 2x = 11$ or – $(3 - 2x) = 11$

To solve the first equation: $3 - 2x = 11$, we must subtract 3 from each side, leaving us with: $-2x = 8$.

Next we divide -2 from each side to get $x$alone: $x = -4$

We now know that one equation has $x = -4$, now we must solve the other equation: $-(3 - 2x) = 11$

We must distribute the – to the parts in the parenthesis: $-3 + 2x = 11$

Next we must add 3 to each side, to give us: $2x = 14$

To get x alone, we divide 2 from each side: $x = 7$

The complete solution for the problem is $x = -4; x = 7$

2.

d. $\boldsymbol{x = 3}$ **or** $\boldsymbol{x = 9}$

The absolute value equation can be solved in one of two ways as: $13 - 2x = 5$ or $-(13 - 2x) = 5$

To solve the first equation: $13 - 2x = 5$, we must subtract 13 from each side, leaving us with: $-2x = -6$.

Next we divide -2 from each side to get $x$alone: $x = 3$

We now know that one equation has $x = 3$, now we must solve the other equation: $-(13 - 2x) = 5$

We must distribute the – to the parts in the parenthesis: $-13 + 2x = 5$

Next we must add 13 to each side, to give us: $2x = 18$

To get x alone, we divide 2 from each side: $x = 9$

Therefore we know that the possible values of x can be $x = 3$or $x = 9$.

3.

b. 7 dozen

We can do this problem algebraically by making c =the amount of dozen cinnamon rolls bought and d =the amount of dozen donuts bought

We can create the problem: $10c + 7d = 161$ to show the total amount of money spent on the pastries

Also we can make another algebra equation for this word problem: $c + d = 20$ to show the total dozen of pastries bought. This equation can be substituted into the first to find how many cinnamon rolls were bought. We can do this by substituting for d(donuts). This equation would be: $d = 20 - c$

Now solve for c: $10c + 7(20 - c) = 161$

Distribute 7 to the parenthesis:$10c + 140 - 7c = 161$

Combine the variables together: $3c + 140 = 161$

Subtract 140 from each side: $3c = 21$

Divide 3 from each side to determine the value of c: $c = 7$

Shane purchased 7 dozen cinnamon rolls and 13 dozen donuts.

4.

c. (2,1,3)

$x + 2y - z = 1$ (a)

$2x - y + 2z = 9$ (b)

$x - 2y - 3z = -9$ (c)

To solve this equation, we need to get one of the variables alone to begin to solve the system of solutions, looking over the three equations we can eliminate the variable y from equation (a) and (c) if we add them together.

$x + 2y - z = 1$

$\underline{x - 2y - 3z = -9}$

$2x - 4z = -8$ (d)

Now we need to eliminate y from equation (b), we can do this by multiplying equation (b) by 2 and adding it to equation (a)

$x + 2y - z = 1$

$\underline{4x - 2y + 4z = 18}$

$5x + 3z = 19$ (e)

After eliminating y from the equations, we can look at equations (d) and (e) to see which variable would be easier to eliminate and variable to solve. In this case since one of the z variables has a -, it will be easier to eliminate z and solve for x. To eliminate z, we will need to multiply each equation to reach a number that can eliminate z.

Equation (d) can be multiplied by 3 and equation (e) can be multiplied by 4 to get the common factor of 12.

$3(2x - 4z = -8) \rightarrow 6x - 12z = -24$

$\underline{4(5x + 3x = 19) \rightarrow 20x + 12z = 76}$

$26x = 52$

Solve for x: $x = 2$

Now that we have found x, we can solve for z using equation (d) or (e). We will use equation (d) since the numbers are lower.

$2(2) - 4z = -8$

$4 - 4z = -8$

$-4z = -12$

$z = 3$

We have determined that $x = 2$ and $z = 3$, now we can solve for y using one of our original equations. In this instance, we will use equation (a).

$2 + 2y - 3 = 1$

$-1 + 2y = 1$

$2y = 2$

$y = 1$

The solution set is (2,1,3).

5.

d. 12

We can do this problem algebraically by making t =the amount of teddy bears bought and s =the amount of sock monkeys bought

We can create the problem: $15t + 12s = 396$ to show the total amount of money spent on the toys

Also we can make another algebra equation for this word problem: $t + s = 30$ to show the total toys bought. This equation can be substituted into the first to find how many teddy bears were bought. We can do this by substituting for s(sock monkeys). This equation would be: $s = 30 - t$

Now solve for t: $15t + 12(30 - t) = 396$

Distribute 12 to the parenthesis:$15t + 360 - 12c = 396$

Combine the variables together: $3t + 360 = 396$

Subtract 360 from each side: $3t = 36$

Divide 3 from each side to determine the value of t: $t = 12$

Gillian purchased 12 teddy bears and 18 sock monkeys.

6.

a. $y > -1$ and $y < x + 2$

The two lines on the graph are best represented by $y > -1$ and $y < x + 2$. The line is drawn as a dashed line to show the points on the line do not satisfy the inequality. The region we wish to find is above the $y > -1$ line and to the right of the $y < x + 2$.

7.

b. (3,1)

Out of the four answer choices, only (3,1) are the only point that lie in the solution set for the system to make it true as shown:

$(1) - 2(3) < 5$

$1 - 6 < 5$

$-5 < 5$ (True)

$4(1) + 3(3) \geq 10$

$4 + 9 \geq 10$

$13 \geq 10$ (True)

8.

d. $\begin{cases} y \leq \frac{1}{2}x + 2 \\ y \geq -x + 2 \end{cases}$

The two lines on the graph are best represented by $y \leq \frac{1}{2}x + 2$ and $y \geq -x + 2$. The line is drawn as a solid line to show the points on the line do satisfy or equal the inequality.

9.

d. $\mathbf{6x^3 + x^2 - 6 + 9}$

To solve this problem, each term of the expression is multiplied by every term of the other expression. Then like terms in the product are added together. You can use a vertical or horizontal format to find the product. Either format is acceptable to use. In this equation, we will use the vertical format:

$2x + 3$

$\underline{3x^2 - 4x + 3}$

$6x^3 + 9x^2$

$-8x^2 - 12x$

$\underline{+6x + 9}$

Combine like terms together to get the final product: $6x^3 + x^2 - 6x + 9$

10.

c. $\mathbf{4x^2 - 10x + 34 + \frac{104}{x-3}}$

For each term we would take out the number that when multiplied would give us the term that we are dividing by as shown below:

$$
\begin{array}{r l}
 & 4x^2 + 10x + 34 + \frac{104}{x-3} \\
x - 3 & \overline{)4x^3 - 2x^2 + 4x + 2}
\end{array}
$$

$\underline{4x^3 - 12x^2}$ Change signs and add

$10x^2 + 4x$

$\underline{10x^2 - 30x}$ Change signs and add

$34x + 2$

$\underline{34x - 102}$ Change signs and add

104

Therefore, $\frac{4x^3-2x^2+4x+2}{x-3} = 4x^2 + 10x + 34 + \frac{104}{x-3}$

11.

d. $\boldsymbol{-8x^2 + 13x - 1}$

To solve this equation you simply distribute the -2 to the equation then combine like terms as shown:

$(-4x^2 + 7x + 3) - 2(2x^2 - 3x + 2)$

$-4x^2 + 7x + 3 - 4x^2 + 6x - 4$

$-8x^2 + 13x - 1$

12.

c. $\boldsymbol{25y^3 - 15y^2 + 15y - 9}$

To solve this problem, you simply use the FOIL method as shown below:

$(5y^2 + 3)(5y - 3)$

$25y^3 - 15y^2 + 15y - 9$

Since there aren't any like terms the equation remains as $25y^3 - 15y^2 + 15y - 9$

13.

b. $\boldsymbol{x^3 + 14x^2 + 53x + 40}$

To find the volume we must remember the rectangular solid which is: Volume$= l \times w \times h$

Our equation would be: $(x + 5)(x + 1)(x + 8)$

Now we must multiply as follows: $x^2 + x + 5x + 5(x + 8)$

Combine like terms before multiplying $(x + 8)$: $x^2 + 6x + 5$

Now multiply $(x^2 + 6x + 5)$ with $(x + 8)$: $(x^2 + 6x + 5)(x + 8)$

$$x^3 + 8x^2 + 6x^2 + 48x + 5x + 40$$

Combine like terms to get our final answer: $x^3 + 14x^2 + 53x + 40$

14.

a. $\boldsymbol{(3a + c)(9a^2 - 3ac + c^2)}$

To find the equation to the answer we must work out the various possibilities. A simple way to do this is by checking the choices given and multiplying them out. With the equation $(3a + c)(9a^2 - 3ac + c^2)$ the work is shown below:

$27a^3 - 9a^2c + 3ac^2 + 9a^2c - 3ac^2 + c^3$

Then combine like terms to get your answer: $27a^3 + c^3$

The a^2c and ac^2 variables cancels itself out leaving the answer you see above.

15.

b. $\boldsymbol{(3x^2 - 6y)(3x^2 + 6y)}$

The factors that represent the total area of this rectangle must have the xy factor cancel each other out when using the FOIL method to solve the area. Only answers a and d do that, we must also consider that the total area includes a x^4. This can only be achieved by having x^2 times x^2 thus leading us to answer choice a.

This also can be factored to check to see if we are correct:

$(3x^2 - 6y)(3x^2 + 6y)$

$9x^4 + 18x^2y - 18x^2y - 36y^2$

The x^2y variable cancels itself out leaving the total area of $9x^4 - 36y^2$

16.

d. $\mathbf{2(3x + 5)(3x - 5)}$

The only choice to show the complete factorization of $18x^2 - 50$ is $2(3x + 5)(3x - 5)$. To check to see if this is correct we can factor $2(3x + 5)(3x - 5)$ below:

$2(9x^2 - 15x + 15x - 25)$

$2(9x^2 - 25)$

$18x^2 - 50$

The complete factorization does check out.

17.

d. $\mathbf{\frac{x^2-x-6}{x^2-9}}$

To arrive at this answer, we must work through the rational expression:

$\frac{x+2}{x+3} + \frac{6}{x^2-9}$

$\frac{x+2}{x+3} + \frac{6}{(x-3)(x+2)}$

$\frac{x+2(x-3)}{x+3(x-3)} + \frac{6}{(x+3)(x-3)}$

$\frac{x^2-3x+2x-6}{x+3(x-3)} + \frac{6}{(x+3)(x-3)}$

The bottom is the same so it can be multiplied once using the FOIL method and then we can combine like terms.

$\frac{x^2-3x+2x-6+6}{x^2-3x+3x-9}$

The $3x$'s cancel each other out leaving us with:

$\frac{x^2-x-6}{x^2-9}$

18.

a. $\mathbf{-3(x + 3)}$

First we need to find the factors for this expression and cancel out any common factors. Our expression is: $\frac{3x^2-27}{3-x}$

We can pull out the 3 from $3x^2 - 27$ to get $3(x^2 - 9)$

Next we can factor $3(x^2 - 9)$: $3(x + 3)(x - 3)$

Our expression now is: $\frac{3(x+3)(x-3)}{3-x}$

We need to eliminate our denominator by taking -1 out of $x - 3$, this will give us: $\frac{-3(x+3)(-x+3)}{3-x}$

Switch the order of $-x + 3$ to cancel out $3 - x$ on the bottom: $\frac{-3(x+3)(3-x)}{3-x}$

Now we are left with $-3(x + 3)$

19.

c. x

To find the simplest form of this equation we first need to find the factors of the equation

$\frac{x^2+3x}{x+7} \cdot \frac{x^2+5x-14}{x^2+x-6}$

$\frac{x(x+3)}{x+7} \cdot \frac{(x+7)(x-2)}{(x+3)(x-2)}$

Now cancel out variables that are the same to receive our final answer: x

20.

d. Point D

The real part of the equation is 2 and the imaginary part of the equation is 3, which means that on the complex plane, the point on the graph is (2, 3).

21.

b. -1

Since $\sqrt{-1}$ times $\sqrt{-1}$ equals -1, then ii equals -1 and i^2 equals -1 also.

22.

c. $4 - 3i$

The real part of the equation is 4 and the imaginary part of the equation is -3, which means that on the complex plane, the point on the graph is (4,-3). Therefore the equation $4 - 3i$.

23.

a. -6

$2i(3i)$ can be simplified by multiplying the real numbers then the imaginary numbers. 2 multiplied by 3 is 6 and ii equals -1. 6 times -1 equals -6.

24.

c. $3 - i$

To find the equivalent form of $\frac{6}{1+i}$ we need to multiply the complex number to the numerator to eliminate it from the denominator with the opposite of $1 + i$, $1{-}i$. This is shown below:

$$\frac{6}{1+i} \cdot \frac{1-i}{1-i}$$

The numerator becomes $6 - i$, while the denominator becomes: $1 - i + i - i^2$.

The i cancel each other out leaving $1 - i^2$. i^2 is -1 and since there is a negative in front of it, the 1 will become positive giving us $1 + 1$.

Adding this will give us 2, the fraction is now: $\frac{6-i}{2}$

The numbers in the fraction can be reduced since both are divisible by 2, giving us the answer of: $\frac{3-i}{1}$ or $3 - i$.

25.

a. 50

When finding the product of complex numbers, you are using the FOIL method to arrive at the answer as shown below:

$(7 + i)(7 - i)$

$49 + i - i - i^2$

The i cancel each other out, leaving: $49 - i^2$

i^2 equals -1 and we can place it into the equation: $49 - (-1)$ or $49 + 1 = 50$

26.

b. $\boldsymbol{x = \frac{1}{5} + \frac{\sqrt{14}}{5} i; x = \frac{1}{5} - \frac{\sqrt{14}}{5} i}$

The equation is written in standard form: $5x^2 - 2x + 3 = 0$

We divide every term by 5 so that the coefficient of x^2 will be 1: $x^2 - \frac{2}{5}x + \frac{3}{5} = 0$

Next we write the parentheses and move the constant term to the right hand side:

$(x^2 - \frac{2}{5}x \quad) = -\frac{3}{5}$

Now we multiply the coefficient of x by $\frac{1}{2}$ and square the product: $\left(-\frac{2}{5} \cdot \frac{1}{2}\right)^2 = \frac{1}{25}$

Then we add $\frac{1}{25}$ to both sides of the equation: $x^2 - \frac{2}{5}x + \frac{1}{25} = -\frac{3}{5} + \frac{1}{25}$

Now we simplify and solve for x: $\left(x - \frac{1}{5}\right)^2 = -\frac{14}{25}$ simplified

$x - \frac{1}{5} = \pm\sqrt{-\frac{14}{25}}$ square root of both sides

$x - \frac{1}{5} = \pm\frac{\sqrt{14}}{5} i$ solved

27.

a. 9

Standard form is $ax + by + c = 0$, where c is a constant. When you complete the square, you first move the constant to one side and the variables with coefficients in front of them to the other.

For this problem, that has already been done for us. The next step is to divide by 2 and square the resulting answer as shown: $-\frac{6}{2} = -3$

$(-3)^2 = 9$

Now add this number to both sides, therefore the answer is 9.

28.

a. $\boldsymbol{x = \frac{1}{3} + \frac{\sqrt{14}}{3} i; x = \frac{1}{3} - \frac{\sqrt{14}}{3} i}$

We first need to get the equation in standard form. This can be done by multiplying both sides by x^2 as shown:

$x^2(3 + \frac{5}{x^2} = \frac{2}{x})$

$3x^2 + 5 = 2x$

$3x^2 - 2x + 5 = 0$

Now that it is in standard form, we can now use the quadratic formula to solve:

$$x = \frac{-(-2) \pm \sqrt{(-2)^2 - 4(3)(5)}}{2(3)} = \frac{2 \pm \sqrt{-56}}{6}$$

The number is divisible by 2 and makes our answer to be:

$x = \frac{1}{3} + \frac{\sqrt{14}}{3}i$ or $x = \frac{1}{3} - \frac{\sqrt{14}}{3}i$

29.

c. One graph has a vertex that is a maximum, while the other graph has a vertex that is a minimum.

The two equations we have are already in the best form to see what they look like.

The vertex form of a parabola (opening up or down) is:

$y = a(x - h)^2 + k$

The vertex of the parabola is (h, k).

If a is positive, the parabola opens upward, and the vertex is a minimum.

If a is negative, the parabola opens downward, and the vertex is a maximum.

Also, the bigger a is, the faster the parabola grows, and the skinnier it will look. The smaller a is, the slower the parabola grows.

We have two parabolas that are almost identical having the same vertices:

$y = -3(x + 2)^2 + 1$ vertex $(2,1), a = -3$

$y = 3(x + 2)^2 + 1$ vertex $(2, 1), a = 3$

The a's are both three but one is negative and one is positive. This means that one graph has a vertex that is a maximum, while the other graph has a vertex that is a minimum making answer choice c as the answer.

30.

b. down 6 and 5 to the left

To move a function up, you add outside the function: $f(x) + b$ is $f(x)$ moved up b units. Moving the function down works the same way; $f(x) - b$ is $f(x)$ moved down b units.

In the first equation we see $y = (x - 3)^2 + 4$

b is 4

In the second equation $y = (x + 2)^2 - 2$ and b is -2

Therefore the b is moving from 4 to -2 on the graph meaning it will move down 6 places

To shift a function left, add inside the function's argument: $f(x + b)$ gives $f(x)$ shifted b units to the left. Shifting to the right works the same way; $f(x - b)$ is $f(x)$ shifted b units to the right.

In our first equation the b is -3 and in the second equation the b is 2.

The common mistake is to think that $f(x + 2)$ moves $f(x)$ to the right by two, because +2 is to the right. But the left-right shifting is backwards from what you might have expected. Adding moves you left; subtracting moves you right.

Therefore it is moving of the graph from 3 to -2 meaning it will move left 5 places.

31.

c. $-\frac{3}{4}$ and 2

Let $y = 0$ to solve for x-intercepts:

$8x^2 - 10x - 12 = 0$

$(4x + 3)(2x - 4) = 0$

$4x + 3 = 0; 4x = -3; x = -\frac{3}{4}$

$2x - 4 = 0; 2x = 4; x = 2$

The x-intercepts are $-\frac{3}{4}$ and 2

32.

b.

Graph b is the best representation of $y = (x + 1)^2 - 3$. To better understand this, we need to take the equation from vertex form to standard form to get a clearer picture of how it would be graphed.

$(x + 1)(x + 1) - 3$

$x^2 + x + x + 1 - 3$

$y = x^2 + 2x - 2$

Our vertices would be $(-1, -3)$ and would be the point that we would see graphed on the parabola. We find the vertices by looking at our original equation the h and k would be our points.

33.

d. (4,3)

To solve for the vertices in a standard form, we must first find x. The equation to find x is $x = -\frac{b}{2a}$

Once we find x, it can be plugged into our equation to find y.

$x = -\frac{-8}{2(1)}$

$x = \frac{8}{2}$

$x = 4$

$y = (4)^2 - 8(4) + 19$

$y = 16 - 32 + 19$

$y = 3$

Our ordered pair for the vertex is $(4,3)$

34.

a. $\boldsymbol{y = \frac{1}{4}x; y = -\frac{1}{4}x}$

The slopes of the two asymptotes will be of the form $m = \pm\frac{a}{b}$.

The number under x is a and the number under y is b.

Therefore our fraction would be$\pm\frac{1}{4}x$

35.

b. hyperbola

The equation given, $\frac{(x-h)^2}{b^2} + \frac{(y-k)^2}{a^2} = 1$ is best represented as a hyperbola with a vertical major axis.

36.

c. $\boldsymbol{\frac{(x+2)^2}{4} - \frac{(y-4)^2}{2} = 1}$

We need to get the problem into standard form. First we need to put similar variables in the problem near to each other: $2x^2 + 8x - 4y^2 - 32y + 12 = 0$

Move 12 to the opposite side: $2x^2 + 8x - 4y^2 - 32y = -12$

Next we need to get the squared variables not to have a number in front:

$2(x^2 + 4x \quad) - 4(y^2 - 8y \quad) = -12$

We must put in the missing number from the square to place in their respected equations and add to the opposite side: $2(x^2 + 4x + 4) - 4(y^2 - 8y + 16) = -12 + 4 + 16$

Factor the double square and add like terms on the opposite side:

$$2(x + 2)^2 - 4(y - 4)^2 = 8$$

Divide 2 and 4 from each side to get 8 on the opposite side to be 1 which will make it into standard form and give us the answer of:

$$\frac{(x + 2)^2}{4} - \frac{(y - 4)^2}{2} = 1$$

37.

d. $\boldsymbol{x = \frac{\log_{10} 19}{\log_{10} 7}}$

The equation $7^x = 19$ is a logarithm. To solve for x, you must divide 7 from each side to get an answer. The answer for this equation is: $x = \frac{\log_{10} 19}{\log_{10} 7}$

38.

c. $\boldsymbol{\frac{1}{10}}$

To find out the value of x, we must divide the $\log_{10}$ from each side and since it cannot be a negative it is written as followed: $\frac{1}{10^1}$

Now solve the fraction: $\frac{1}{10}$

39.

b. $\boldsymbol{4^x = \frac{1}{8}}$

$\log_4 \frac{1}{8} = x$ can be written as the $\log$ raised to the x power equaling whatever number right next to the $\log$. In this case our $\log$ is 4 so it would be: $4^x = \frac{1}{8}$

40.

a. Step 1

The equation $\log_3 \frac{3}{81}$ is shown below on how to be correctly solved:

Step 1: $\log_3 \frac{3}{81} = \log_3 3 - \log_3 81$

Step 2: $= 1 - 4$

Step 3: $= -3$

The equation needed to be subtracted instead of added since the log was in fraction form.

41.

a. Markus

To solve x^2x^{-5} you must take the negative exponent and place it on the bottom to make it positive: $\frac{x^2}{x^5}$

Subtract 5-2 to make the numerator 1 and the denominator the remaining exponent: $\frac{1}{x^3}$

42.

b. Step 2

The first step was worked out correctly. It was not until Step 2 that the first error occurred. Instead of multiplying one side by $x - 3$ and multiplying the other side by -1, it should have been multiplied on both sides to balance out the equation. The problem is done correctly below:

Step 1: $\log_{10}(\frac{x+3}{x-3}) = 7$

Step 2: $\log_{10}(x + 3) = 7x - 21$

Step 3: $24 = 6x$

Step 4: $x = 4$

43.

d. 2000

To solve this problem, simply plug in the numbers given into the equation: $y = 18000\left(\frac{1}{3}\right)^{\frac{120}{60}}$

$\frac{120}{60}$ can be reduced to 2 making the equation now: $y = 18000\left(\frac{1}{3}\right)^2$

$\frac{1}{3} \times \frac{1}{3} = \frac{1}{9}$

$y = 18000\left(\frac{1}{9}\right)$ or $y = \frac{18000}{9}$

$y = 2000$

The model toy car will have only 2000 cars left in 120 days.

44.

d. $y = (300) \cdot (3)^t$

To create an equation for this word problem, we must remember that we need to use a variable that remains as a starting point. The mold begins at 300 bacteria and increases by each day.

From the sample chart we are given, we can see that the number doubles with each day. Therefore we can make the conclusion that our starting point of 300 is multiplied by 3 that has the exponent of t.

45.

c. $\mathbf{\frac{1}{16}}$

To solve for 4^x, we will allow $y = 0$.

$4^x = 0$

$x = \frac{1}{4^2}$

$x = \frac{1}{16}$

Therefore our x-axis would be $\frac{1}{16}$

46.

a. $\mathbf{\frac{\log_{10} 46}{\log_{10} 4}}$

Another way that this logarithm can be written out is by dividing the number that is next to the logarithm by the log base making the log base to be now a $\log_{10}$, with the problem being $\frac{\log_{10} 46}{\log_{10} 4}$

47.

c. 2

The value of $\log_4 16$ can be determined by knowing that 16 is 4^2. Therefore the value is 2.

48.

b. 1.126

To find the approximate value of $\log 196$, we first must determine the 2 and 7 factors of 196.

Factors of 196 using only 2 and 7 are: $2 \cdot 2 \cdot 7 \cdot 7$

Next we add up the values of $\log 2$ and $\log 7$ to get our approximate value of $\log 196$

$$0.178 + 0.178 + 0.385 + 0.385 = 1.126$$

The approximate value of $\log 196$ is 1.126

49.

c. some values of x

The equation $\frac{x^2}{4} = x$ will only work with some number, not all.

For example if $x = 4$, our equation would be $\frac{4^2}{4} = 4$

$\frac{16}{4} = 4$ making the equation true.

However if $x = 3$, our equation would be: $\frac{3^2}{4} = 3$

$\frac{9}{4}$ does not equal 3 therefore the equation would be false.

Thus the equation is true with some values of x.

50.

d. The equation is never true.

The equation is never true when x is equal to any number.

For example when $x = 2$, the equation looks like this: $\frac{(2)^2+25}{2+5} = 2 - 5$

$$\frac{4 + 25}{7} = -3$$

$\frac{29}{7}$ does not equal -3, therefore the equation is not true.

We can use another number to prove this point, we can use $x = 5$ to prove the equation is false.

$$\frac{(5)^2 + 25}{5 + 5} = 5 - 5$$

$$\frac{25 + 25}{10} = 0$$

$$\frac{50}{10} = 0$$

Five does not equal zero therefore when $x = 5$, along with any other number, the equation is false.

51.

b. y is equal to non-negative real numbers

In the equation, $y = \sqrt{x^2}$ all numbers that are multiplied with themselves will be positive, except for zero which is neither positive or negative causing y to be equal to non-negative real numbers.

Three examples are shown below to prove this to be true.

$y = \sqrt{(-1)^2}$	$y = \sqrt{(0)^2}$	$y = \sqrt{(3)^2}$
$y = \sqrt{1}$	$y = \sqrt{0}$	$y = \sqrt{9}$
$y = 1$	$y = 0$	$y = 3$

52.

b. 48

There are 2 different combinations of al that Albert could use (al or la)

There are 24 different combinations of 1987 that Albert could use ($4!$ which is 4 factorial or $4 \cdot 3 \cdot 2 \cdot 1 = 24$)

You multiply both parts $2 \cdot 24$ and get 48.

53.

b. 720

If the solo artist is always first and the vocal duo is always last at the festival, this means that the 6 bands can be placed in any order. To find this number, we can use $6!$ (which is 6 factorial)

$$6 \cdot 5 \cdot 4 \cdot 3 \cdot 2 \cdot 1 = 720$$

54.

a. $\frac{1}{910}$

The first recipient can be chosen in 15 ways
The second recipient can be chosen in 14 ways, after the first person is selected.
The third recipient can be chosen in 13 ways, after the first and second person is selected.

So there are $15 \cdot 14 \cdot 13 = 2730$ different choices for first and second choices.

In this matter, the order is not important. There are twice as many options as necessary; therefore we need to divide by 3. There are a total of 2730 groups of 3 recipients. One of those groups will contain Christopher, Justin, and Ryan, therefore $\frac{1}{910}$ is the probability of all 3 receiving the scholarship.

55.

c. $16y^4 - 96y^3 + 216y^2 - 216y + 81$

Below is the expanded binomial expression and the steps to arrive at our answer:

$(2y - 3)^4 = (2y - 3)(2y - 3)(2y - 3)$

$(4y^2 - 12y + 9)(4y^2 - 12y + 9)$

$16y^4 - 48y^3 + 36y^2 - 48y^3 + 144y^2 - 108y + 36y^2 - 108y + 81$

Combine like terms to get the answer: $16y^4 - 96y^3 + 216y^2 - 216y + 81$

56.

d. 31

In a linear equation (a binomial to the first power) there are two terms. For a cubic (binomial cubed) there are 4 terms. There is a pattern that emerges from this: a binomial expansion of $(a + b)^n$ will have $n + 1$ terms.

In the problem given $(2x^4 + y^3)^{30}$ our binomial expansion contains 31 terms.

57.

c. $x^4 + 4x^3y + 6x^2y^2 + 4xy^3 + y^4$

The simplest way to find the expanded binomial when we have a $(x + y)^n$ or a larger exponent is to use Pascal's triangle.

The equation is raised to the 4th power which means that there will be 5 terms. You would then look at the triangle to determine the number that will be in from of each expression. For this problem, you would look at the 5th row and find the numbers: 1,4,6,4,1

The exponents for each coefficient would go in decreasing order for x and increased order for y, giving us answer choice c.

58.

a. Yes, all steps are correct.

The steps that were written by Paul for the mathematical induction $3^n - 1$ is a multiple of 2 is correct.

59.

d. 88

To find the 14th number in the arithmetic series we must use the formula $a_n = a_1 + (n-1)d$

Our common difference is 6 making $d = 6$

The number we are wanting to find is the 14th term, this is n.

a_1 is the first number in the sequence which is 4

Our equation: $a_{14} = 4 + (14-1)6$

$4 + (13)6$

$4 + 84$

$a_{14} = 88$

60.

c. 5

To find the common ratio, we have to divide a pair of terms. It does not matter which pair is picked, as long as they are right next to each other:

$$\frac{10}{2} = 5; \frac{50}{10} = 5; \frac{250}{50} = 5$$

Each number that is right next to each other has the same common ratio of 5.

61.

a.15

To solve this problem you must place $f(x)$ equation into the x part of the $g(x)$ equation and the x in the $f(x)$ equation equals 1.

This is shown below:

$2(4(1)^2 + 5) - 3$

$2(4(1) + 5) - 3$

$2(4 + 5) - 3$

$2(9) - 3$

$18 - 3 = 15$

62.

a. $\boldsymbol{x^2 + 4x + 1}$

We need to replace the x in $f(x)$ with the entire expression of $g(x)$ as shown below:

$$(x + 2)^2 - 3$$

Now we will factor and combine like terms to get our answer:

$(x + 2)(x + 2) - 3$

$x^2 + 2x + 2x + 4 - 3$

$x^2 + 4x + 1$

63.

a. $\boldsymbol{5x^2 - 11x + 22}$

To find the sum of $f(x) + g(x)$, first we must factor out the equation $g(x)$ before we can combine like terms.

$4(x - 2)^2$

$4(x - 2)(x - 2)$

$4(x^2 - 4x + 4)$

Multiply $4x^2 - 16x + 16$

Add $f(x)$ and $g(x)$: $4x^2 - 16x + 16 + x^2 + 5x + 6$

$5x^2 - 11x + 22$

64.

b. 24%

If the chance of snow in Nashville = 20%, then the chance that it won't rain is 80% or $\frac{4}{5}$.
If the chance of snow in Buffalo = 70%, then the chance that it won't snow is 30% or $\frac{3}{10}$.
Multiplying $\frac{4}{5} \cdot \frac{3}{10} = \frac{12}{50} = .24 \text{ or } 24\%$

65.

b. $\frac{7}{16}$

7 large blue + 9 large green = 16 large circles
3 small red + 1 small blue = 4 small circles but we're only concerned with the 16 large circles.
The odds of a large circle being blue = $\frac{part}{whole}$

$\frac{number\ of\ large\ blue\ circles}{number\ of\ large\ circles}$

$= \frac{7}{16}$

66.

d. The standard deviation will not change as the mean will increase.

The standard deviation is a measurement of how far numbers are away from the mean. So if you add 4 to each number, the mean will also go up by 4, therefore the numbers will be the same distance from the mean as they were originally, so the standard deviation remains the same.

Question Number	Correct Answer	Standard
1	A	1.0
2	D	1.0
3	B	2.0
4	C	2.0
5	D	2.0
6	A	2.0
7	B	2.0
8	D	2.0
9	D	3.0
10	C	3.0
11	D	3.0
12	C	3.0
13	B	3.0
14	A	4.0
15	B	4.0
16	D	4.0
17	D	7.0
18	A	7.0
19	C	7.0
20	D	5.0
21	B	5.0
22	C	5.0
23	A	6.0
24	C	6.0
25	A	6.0
26	B	8.0
27	A	8.0
28	A	8.0
29	C	9.0
30	B	9.0
31	C	10.0
32	B	10.0
33	D	10.0
34	A	16.0
35	B	16.0
36	C	17.0
37	D	11.1
38	C	11.1
39	B	11.1
40	A	11.2
41	A	11.2

42	B	11.2
43	D	12.0
44	D	12.0
45	C	12.0
46	A	13.0
47	C	14.0
48	B	14.0
49	C	15.0
50	D	15.0
51	B	15.0
52	B	18.0
53	B	18.0
54	A	19.0
55	C	20.0
56	D	20.0
57	C	20.0
58	A	21.0
59	D	22.0
60	C	22.0
61	A	24.0
62	A	24.0
63	A	25.0
64	B	PS1.0
65	B	PS2.0
66	D	PS7.0

Made in the USA
San Bernardino, CA
28 September 2015